THE FRONTIERS COLLECTION

THE FRONTIERS COLLECTION

Series Editors:
A.C. Elitzur M.P. Silverman J. Tuszynski R. Vaas H.D. Zeh

The books in this collection are devoted to challenging and open problems at the forefront of modern science, including related philosophical debates. In contrast to typical research monographs, however, they strive to present their topics in a manner accessible also to scientifically literate non-specialists wishing to gain insight into the deeper implications and fascinating questions involved. Taken as a whole, the series reflects the need for a fundamental and interdisciplinary approach to modern science. Furthermore, it is intended to encourage active scientists in all areas to ponder over important and perhaps controversial issues beyond their own speciality. Extending from quantum physics and relativity to entropy, consciousness and complex systems – the Frontiers Collection will inspire readers to push back the frontiers of their own knowledge.

H. Dieter Zeh

THE PHYSICAL BASIS OF
THE DIRECTION
OF TIME

5th edition

With 35 Figures

 Springer

Professor Dr. H. Dieter Zeh
University of Heidelberg,
Institute of Theoretical Physics,
Philosophenweg 19,
69120 Heidelberg, Germany
email: zeh@uni-heidelberg.de

Series Editors:

Avshalom C. Elitzur
Bar-Ilan University,
Unit of Interdisciplinary Studies,
52900 Ramat-Gan, Israel
email: avshalom.elitzur@weizmann.ac.il

Rüdiger Vaas
University of Gießen,
Center for Philosophy and Foundations of Science
35394 Gießen, Germany
email: Ruediger.Vaas@t-online.de

Mark P. Silverman
Department of Physics, Trinity College,
Hartford, CT 06106, USA
email: mark.silverman@trincoll.edu

H. Dieter Zeh
University of Heidelberg,
Institute of Theoretical Physics,
Philosophenweg 19,
69120 Heidelberg, Germany
email: zeh@uni-heidelberg.de

Jack Tuszynski
University of Alberta,
Department of Physics, Edmonton, AB,
T6G 2J1, Canada
email: jtus@phys.ualberta.ca

Cover figure: Image courtesy of the Scientific Computing and Imaging Institute,
University of Utah (www.sci.utah.edu).

ISSN 1612-3018
ISBN 978-3-642-08760-8   e-ISBN 978-3-540-68001-7

Springer is a part of Springer Science+Business Media

springer.com

© Springer-Verlag Berlin Heidelberg 2007
Softcover reprint of the hardcover 5th edition 2007

The use of general descriptive names, registered names, trademarks, etc. in this publication does not imply, even in the absence of a specific statement, that such names are exempt from the relevant protective laws and regulations and therefore free for general use.

Cover design: KünkelLopka, Werbeagentur GmbH, Heidelberg

Preface

Four previous editions of this book were published in 1989, 1992, 1999, and 2001. They were preceded by a German version (Zeh 1984) that was based on lectures I had given at the University of Heidelberg.

My interest in this subject arose originally from the endeavor to better understand all aspects of irreversibility that might be relevant for the statistical nature and interpretation of quantum theory. The quantum measurement process is often claimed to represent an 'amplification' of microscopic properties to the macroscopic scale in close analogy to the origin of classical fluctuations, which may lead to the local onset of a phase transition, for example. This claim can hardly be upheld under the assumption of universal unitary dynamics, as is well known from the example of Schrödinger's cat. However, the classical theory of statistical mechanics offers many problems and misinterpretations of its own, which are in turn related to the oft-debated retardation of radiation, irreversible black holes with their thermodynamical aspects, and – last but not least – the expansion of the Universe. So the subject offered a great and exciting 'interdisciplinary' challenge. My interest was also stimulated by Paul Davies' (1977) book that I used successfully for my early lectures. Quantum gravity, that for consistency has to be taken into account in cosmology, even requires a complete revision of the concept of time, which leads to entirely novel and fundamental questions of interpretation (Sect. 6.2).

Many of these interesting fields and applications have seen considerable progress since the last edition came out. So, while all chapters have again been thoroughly revised for this fifth edition in order to take these developments into account, changes concentrate on Sects. 2.3 (Radiation Damping), 4.3 (Decoherence), 4.6 (Interpretations of Quantum Theory), 5.3 (Expansion of the Universe) and Chap. 6 (Quantum Cosmology). There are new Sects. 3.5 (on Cosmic Probabilities and History) and 4.3.3 (on Quantum Computers), while Sect. 5.3 has been subdivided and extended. In general, I have tried to remove 'vague' statements, or to make them more precise – although this was not always possible because of the complexity or even speculative nature of some fields. As in previous editions, the focus of the book is on questions of interpretation and relations between different fields – not on technical formalisms and empirically unfounded or predominantly mathematical ideas and concepts.

Many friends and colleagues helped me with their advice on various subjects during the preparation of all previous editions. I cannot here repeat all their names (I hope they are all duly mentioned in the corresponding previous prefaces), but I wish to thank here my former collaborators Erich Joos and Claus Kiefer for their enduring support to all editions. Special thanks this time go to Angela Lahee for her encouragement to prepare a fifth edition (the first one for the Springer Frontiers Collection), and to Stephen Lyle for editing it (although he is not responsible for any errors I may have introduced with numerous last-minute corrections).

I intend to post corrections or revisions to some sections of the book at my website www.time-direction.de whenever it should turn out to be appropriate.

Heidelberg,   H.D. Zeh
April 2007

Contents

Introduction

The asymmetry of Nature under a 'reversal of time' (that is, a reversal of motion and change) appears only too obvious, as it deeply affects our own form of existence. If physics is to justify the hypothesis that its laws control everything that happens in Nature, it should be able to explain (or consistently describe) this fundamental asymmetry which defines what may be called a *direction in time* or even – as will have to be discussed – a direction *of* time. Surprisingly, the very laws of Nature are in pronounced contrast to this fundamental asymmetry: they are essentially symmetric under time reversal. It is this discrepancy that defines the enigma of the direction of time, while there is no lack of asymmetric dynamical formalisms or pictures that go beyond the fundamental empirical laws.

It has indeed proven appropriate to divide the formal dynamical description of Nature into *laws* and *initial conditions*. Wigner (1972), in his Nobel Prize lecture, called this conceptual distinction Newton's greatest discovery, since it demonstrates that the laws by themselves are far from determining Nature. The formulation of these two pieces of the dynamical description requires that appropriate *kinematical* concepts (formal *states* or *configurations* z, say), which allow the unique mapping (or 'representation') of all possible states of physical systems, have already been defined on empirical grounds.

For example consider the mechanics of N *mass points*. Each state z is then equivalent to N points in three-dimensional space, which may be represented in turn by their $3N$ coordinates with respect to a certain frame of reference. States of physical *fields* are instead described by certain functions on three-dimensional space. If the laws of Nature contain kinematical elements (constraints on kinematical concepts that would otherwise be too general, such as $\mathrm{div}\,\boldsymbol{B} = 0$ in electrodynamics), one should distinguish them from the dynamical laws proper. This is only in *formal* contrast to relativistic spacetime symmetry (see Sect. 5.4).

The laws of Nature, thus refined to their purely dynamical content, describe the time dependence of physical states, $z(t)$, in a general form – usually by means of differential equations. They are called *deterministic* if they uniquely determine the state at time t from that (and possibly its time derivative) at any earlier or later time, that is, from an appropriate initial or final condition. This symmetric dynamical determinism is much more rigorous than the traditional concept of *causality*, which requires that every event in Nature must possess a specific *cause* (in its past), while not necessarily a specific *effect* (in its future). The *Principle of Sufficient Reason* (or at least its 'causal root'[1]) can be understood in this asymmetric causal sense that would define an absolute direction of time.

However, only since Newton has uniform motion been interpreted as 'eventless' (thus not needing a cause), while acceleration requires a *force* as the modern form of *causa movens* (usually assumed to act in a retarded, but hardly ever in an advanced manner). From the ancient point of view, terrestrial bodies were usually regarded as eventless or 'natural' only when at rest, and celestial ones when moving in circular orbits (later also including epicycles), or when at rest on the celestial ('crystal') spheres. These motions thus did not require any dynamical causes according to this picture, similar to uniform motion today. None of the traditional causes (neither physical nor others) ever questioned the fundamental asymmetry in (or of) time, since there were no conflicting symmetric dynamical laws yet.

Newton's concept of a force determining acceleration (the *second* time derivative of the 'state') forms the basis of the formal Hamiltonian concept of states in *phase space* (with corresponding dynamical equations of *first* order in time). First order time derivatives of states in configuration space, required to define momenta, can then be freely chosen as part of the initial conditions. In its Hamiltonian form, this part of the kinematics is intermingled with dynamics, as the definition of canonical momentum depends in general on a dynamical concept (the Lagrangean).

Newton recognized friction as a source of time asymmetry. While different motions which may start from one and the same unstable position of rest would require different initial perturbations as sufficient reasons, friction (if understood as a fundamental force) could deterministically bring different motions to the same rest. States at which the symmetry of determinism may thus come to an end (perhaps asymptotically in time) are called *attractors* in some theories.

The term 'causality' is unfortunately understood in very different ways. In physics, it is often synonymous with dynamical determinism, or it may refer to the relativistic limits for the propagation of causal effects, based on the light cone structure of spacetime. In philosophy, it sometimes means the existence of laws of Nature in general. In mathematical physics, dynamical determinism is often understood asymmetrically as applying only in the 'forward' direc-

---

[1] Its 'logical root' has nothing to do with time, but is often confused with dynamical causality. For example, logical *operations* are performed in time by physical systems, even though they can, in a strict sense, only lead to tautologies, which are true regardless of any physical operations (see also the end of Sect. 3.3).

tion of time (thus allowing attractors – see Sect. 3.4). Time-reversal-symmetric determinism was discovered together with the dynamical laws of mechanics in situations where friction could be neglected (as in celestial mechanics). An asymmetric concept of 'intuitive causality' that is compatible with (though essentially different from) symmetric determinism will be defined and discussed in the introduction to Chap. 2.

A subtle but important point here is that the time reversal symmetry of the *concept of determinism* does not necessarily require symmetric dynamical laws. For example, the Lorentz force $e\boldsymbol{v} \times \boldsymbol{B}$, acting on a charged particle and resulting from a *given* external magnetic field, changes sign under time reversal (defined by replacing t by $-t$).[2] Nonetheless, determinism applies equally in both directions of time. This is possible, since the time reversal asymmetry of this equation of motion may be compensated for by a simultaneous reversal of the magnetic field.

Other (more or less physical) *compensating symmetry operations* are known (see Atkinson 2006). For example, the time reversal symmetry of the Schrödinger equation is restored by complex conjugation of the wave function. This can be described by means of Wigner's anti-unitary operation T which leaves the configuration basis unchanged, $Tc|q\rangle = c^*|q\rangle$ for complex numbers c. T may be chosen to contain further self-inverse operations, such as multiplication with the matrix β for the Dirac equation. A trivial example is the time reversal of *states in classical phase space*, $q, p \to q, -p$. This transformation restores invariance of the Hamiltonian equations, which would be violated under a *formal* time reversal $p(t), q(t) \to p(-t), q(-t)$. In quantum theory it corresponds to the transformation $T|p\rangle = |-p\rangle$, which is now a *consequence* of anti-unitarity when T is applied to the state $|p\rangle = (2\pi)^{-1/2} \int \mathrm{d}q\, \mathrm{e}^{ipq}|q\rangle$.

For trajectories of states, $z(t)$, one usually includes the transformation $t \to -t$ in the action of T rather than applying the latter only to the state z: $Tz(t) := z_T(-t)$, where $z_T := Tz$ is the 'time-reversed state' defined above. In the Schrödinger picture of quantum theory this is again automatically taken care of by the anti-unitarity of T when commuted with the time translation e^{-iHt} for a time reversal invariant Hamiltonian H. In this sense, 'T invariance' does indeed mean time reversal invariance. When discussing time reversal, one usually also presumes invariance under *translations* in time in order not to specify an arbitrary origin for the time reversal transformation $t \to -t$.

The time reversal asymmetry characterizing weak forces, which is responsible for K-meson decay, may similarly be compensated for by an additional CP transformation, where C and P are charge conjugation and spatial reflection, respectively. The latter do *not* just reflect a time reversal elsewhere

---

[2] Any distinction between reversal of time and reversal of motion (or any other kind of change) would require some concept of absolute or extraphysical time (see Chap. 1). For example, an asymmetry of the fundamental dynamical laws would *define* (or presume) an absolute direction of time – just as Newton's equations define absolute time up to linear transformations (which thus do allow a reversal of sign).

(such as the inversion of a magnetic field that is caused by the reversal of external currents). Only if the compensating symmetry transformation represents an observable, such as CP, and is not the consequence of a time reversal elsewhere, does one speak of a *violation* of time reversal invariance of the dynamics.

The possibility of compensating for a dynamical time reversal by another symmetry transformation (observable or not) reflects the prevailing *symmetry of determinism*. Such 'symmetric' violations of time reversal invariance have therefore nothing to do with irreversibility, which forms the subject of this book. All known *fundamental* laws of Nature are symmetric under time reversal after compensation by an appropriate symmetry transformation, thus defining a combined symmetry, say $\hat{T}$. For example, $\hat{T} = CPT$ in particle physics, while $\hat{T}\{\boldsymbol{E}(\boldsymbol{r}), \boldsymbol{B}(\boldsymbol{r})\} = \{\boldsymbol{E}(\boldsymbol{r}), -\boldsymbol{B}(\boldsymbol{r})\}$ in classical electrodynamics. This means that for every solution $z(t)$ of the dynamical laws there is precisely one time-reversed solution, $z_{\hat{T}}(-t)$, where $z_{\hat{T}} = \hat{T}z$. This fact is essential for all statistical arguments regarding irreversibility.

'Initial' conditions are usually understood as conditions which fix the integration constants, that is, which select *particular* (individual) solutions of the equations of motion. They could just as well be formulated as final conditions, even though this would not represent the usual *operational* (hence asymmetric) application of the theory. These initial conditions may select the solutions which are 'actually' found in Nature. An individual (contingent) trajectory $z(t)$ is generically *not* symmetric under time reversal, that is, not *identical* with $z_{\hat{T}}(-t)$. If $z(t)$ is sufficiently complex, the time-reversed process is not even likely to occur anywhere else in Nature within reasonable approximation.

However, most phenomena observed in Nature violate time reversal symmetry in a less trivial way if considered as whole *classes of phenomena*. The members of some class may be abundant, while the time-reversed class is not realized at all. Such symmetry violations will be referred to as 'fact-like' – in contrast to the mentioned CP symmetry violations, which are 'law-like'. In modern versions of quantum field theory, even the boundary between laws of Nature and initial conditions gets blurred. Certain parameters which are usually regarded as part of the laws (such as those characterizing the mentioned CP violation) may have arisen by *spontaneous symmetry breaking* (an indeterministic irreversible process of disputed nature in quantum theory – see Sects. 4.6 and 6.1).

In contrast to what is often claimed in textbooks, this asymmetric appearance of Nature cannot be explained by statistical arguments. If the laws are invariant under time reversal when compensated by another symmetry transformation, there must be precisely as many solutions in the time-reversed class as in the original one (see Chap. 3).

Since Eddington, classes of phenomena characterizing a direction in time have been called *arrows of time*. The most important ones are:

1. **Radiation.** In most situations, fields interacting with local sources are appropriately described by means of retarded (outgoing or defocusing) solutions (see Chap. 2). A spherical outgoing wave is observed *after* a point-like event that represents a source. This may lead to *damping* of the source. One may easily observe 'spontaneous' emission (when incoming radiation is absent – see Item 5), while absorption requires retarded consequences. Even an ideal absorber leads to *retarded* shadows (destructive interference with a retarded field).

2. **Thermodynamics.** The Second Law $dS/dt \geq 0$ is often regarded as a *law* of Nature. In microscopic description it has instead to be interpreted as *fact-like* (Chap. 3). This arrow is certainly the most general and important one. Because of its applicability to human memory and other physiological processes, it may also be responsible for the *impression* that time itself has a direction (corresponding to an *apparent flow of time* – see Chap. 1).

3. **Evolution.** Dynamical 'self-organization' of matter, for example observed in biological and social evolution, may appear to contradict the Second Law. However, *global* entropy always keeps growing if the environment is properly taken into account (Sect. 3.4).

4. **Quantum Mechanical Measurement.** The probability interpretation of quantum mechanics is usually understood as describing a fundamental indeterminism of the future, although its interpretation and compatibility with the deterministic Schrödinger equation constitutes a long-standing open problem. Stochastic quantum 'events' are often dynamically described by a collapse of the wave function – not only in measurements. The Schrödinger equation itself may describe growing entanglement as an arrow of time that is analogous to (but different from) statistical mechanics – see Chap. 4.

5. **Exponential Decay.** Many unstable physical systems *decay* exponentially in time (see Sect. 4.5). Exponential growth is only observed under specific circumstances in self-organizing systems (Item 3 above).

6. **Gravity** seems to 'force' massive objects to move towards each other with increasing time. Stars or star clusters contract. However, this is another prejudice about the *causal* (time-directed) action of forces. Gravity describes attraction in *both* directions of time, since Newton's laws are of *second* order. The asymmetry occurs since we often *prepare* objects in an initial state of rest, while the observed contraction of stars against their internal pressure is controlled by thermodynamic and radiation phenomena. On the other hand, gravitating objects are characterized by a negative heat capacity, and classically even by the capacity to contract without limit in accordance with the Second Law if appropriately prepared (see Chap. 5). In general relativity this leads to time asymmetric *future horizons* in spacetime (characterizing black holes), through which objects can only *dis*appear. *Expansion* against gravity is observed for the Universe as a whole – thus indicating an unconventional *cosmic* initial condition. Since cosmic expansion does not define a *class* of phenomena, it has often been

suggested to represent the 'master arrow' from which all others may be derived (see Sects. 5.3 and 6.2).

In spite of their fact-like nature, these arrows of time, in particular the thermodynamical one, have been regarded by some of the most eminent physicists as even more fundamental than the dynamical laws themselves. For example, Eddington (1928) wrote:

> The law that entropy always increases holds, I think, the supreme position among the laws of Nature. If someone points out to you that your pet theory of the Universe is in disagreement with Maxwell's equations – then so much the worse for Maxwell's equations. ... but if your theory is found to be against the second law of thermodynamics, I can give you no hope; there is nothing for it but to collapse in deepest humiliation.

And Einstein (1949) remarked:

> It [thermodynamics] is the only physical theory of universal content concerning which I am convinced that, within the framework of the applicability of its basic concepts, it will never be overthrown.

These remarks were hardly meant to express doubts over the *derivability* of the thermodynamical arrow of time by statistical means when using those 'less credible theories' (see Chap. 3). Rather, they are intended to express their authors' conviction in the *invariance* of the derived thermodynamical concepts and laws under modifications and generalizations of these theories. However, this statistical derivation will be shown to require important assumptions about the initial state of the Universe. *If* the Second Law is fact-like in this sense, its violation or reversal must at least be conceivable, and thus cannot be excluded *a priori*.

The arrows of time listed above characterize an asymmetric *history* of the Universe. This history can be conceived of as a whole, comparable to a movie film sitting on the desk, or an ordered stack of picture frames ('states'), without any selection of a present (that is, of a specific 'actual' frame) or an external distinction between beginning and end. This is called the 'block universe view' (see Price 1996). It may be contrasted with the view of an *evolving* history, observed by an *external* movie viewer as a definer of 'absolute' time for the running movie.

It appears questionable whether these different *views* might possess different power in explaining an asymmetry of the (hi)story described by the movie, but they are regarded as basically different in this respect by many philosophers, including also some physicists (Prigogine 1980, von Weizsäcker 1982). The second view is related to the idea that the past is 'fixed', while the future is 'open' and 'does not yet exist'. The asymmetric history is then regarded as the 'outcome' (or the consequence) of this time-directed 'process of coming-into-being'. (The abundance of quotation marks indicates how our language is loaded with prejudice about the flow of time.) The fact that there are documents, such as fossils, only about the past, and that we cannot remember the future,[3] appears as evidence for this so-called 'structure of time' or the 'historical nature' (*Geschichtlichkeit*) of the world.

However, an asymmetry in the stack of movie frames on the desk is defined regardless of any presentation or production of the movie in external time. *Correlations* between an individual movie frame and certain others, which may represent 'documents' about the latter, are properties of the set of frames on the desk. If consistent asymmetric memory relations existed throughout the whole story, an intrinsic observer, who was part of the story, could know its content only at the (intrinsically defined) 'end'. He could nonetheless conceive of a 'potential' complete story even within the story, in particular if he discovered dynamical laws. Existing 'actually' only in a specific frame, he could neither deny nor prove the existence of other frames, although he might 'remember' those frames which represent his intrinsic past (even if the movie were presented backwards). The time he is aware of has to be read from clocks showing up on the picture frames – not from the watch of an external movie viewer or from any frame numbers (see also Chap. 1 and Sect. 5.4). The concept of 'existence' is here evidently used with various meanings, and the debate may easily become one about words. Similarly, within our 'world movie', concepts like *fixed* and *open*, or *actual* versus *potential*, can only be meaningful either as statements about *practical abilities* of predicting and retrodicting, or as statements about dynamical *models*.

The argument that the historical nature of the world be a prerequisite (in the Kantian sense) for the fact that we *can* make experience does not exclude the possibility (or necessity) of explaining it in terms of those laws and concepts that have been distilled from this experience. They may then be hypothetically extrapolated to form a 'world model', whereby the historical nature may even turn out *not* to apply to other spacetime regions (see Sect. 5.3.3).

In classical physics, the Second Law is usually regarded as the physical basis of the historical nature of the world. Its statistical interpretation would then mean that this 'structure of time' (that is, its apparent direction) is merely the consequence of contingent facts which characterize our specific world. For example, one may explain thermodynamically why there are observations, but no 'un-observations' in which initially present information (memory about the future) would disappear by means of a controlled interaction between the observing and observed systems. This un-observation has to be distinguished from the usual process of forgetting, which represents an information loss in the memory device in accordance with an increase of entropy (see Sect. 3.3).

The concept of 'retarded' information would thus arise as a *consequence* of thermodynamics (and not the other way round, as is sometimes claimed). The

---

[3] "It's a bad memory that only works backwards" says the White Queen to Alice.

inconsistency of presuming either an extra-physical concept of information or extra-physical operations has often been discussed by means of *Maxwell's demon*. In particular, the 'free will' of an experimenter should not be misused to *explain* the specific (low entropy) initial conditions that he prepares in his laboratory. If the experimenter (or demon) were not required to obey the thermodynamical laws himself, his actions could readily *create* the thermodynamical arrow of time observed in his experiments. Nonetheless, the possibility that conscious beings require new fundamental laws cannot *a priori* be excluded.

The indeterminism of quantum measurements and other 'quantum events' has often been interpreted as evidence for such an extra-physical concept of (human?) information. This is documented by many statements by important physicists. For example, Heisenberg argued in the spirit of idealism that "a particle trajectory is created only by its observation,"[4] while von Weizsäcker claimed that only "what has been observed exists with certainty."[5] One can similarly understand Bohr's statement: "Only an observed quantum phenomenon is a phenomenon." He insisted that a quantum measurement cannot be analyzed as an objective dynamical process ("there is no quantum world"). A similar view can be found in Pauli's letter to Born (Einstein, Born and Born 1984): "The appearance of a definite position x_0 during an observation ... is then regarded as a creation existing outside the laws of Nature."[6] Born often expressed his satisfaction with quantum mechanics, as he felt that his probability interpretation saved free will from the determinism of classical laws.

The extra-physical time arrow appears in all *operational* formulations of quantum theory, such as those describing probabilistic relations connecting preparations and subsequent measurements – thus restricting quantum theory to laboratory physics performed by humans. Most of these formulations rely on a given (absolute) direction of time. This should then be reflected by the dynamical description of quantum measurements and 'measurement-like processes' even in the block universe picture. The impact of such phenomena (provided they do indeed occur) on the formal physical description should therefore be precisely located.

Much of the *philosophical* debate about time is concerned with language problems, some of them simply arising from the pre-occupied usage of the tenses, particularly for the verb 'to be' (see Smart 1967, or Price 1996). Aristotle's famous (pseudo-)problem regarding the potential truth value of the claim that there will be a sea battle tomorrow survives not only in *Sein und Zeit*, but even in quantum theory in the form of an occasional confusion of logic with dynamics ('logic of time') – see footnote 1. A careful distinction

---

[4] *"Die Bahn entsteht erst dadurch, daß wir sie beobachten."*

[5] *"Was beobachtet worden ist, existiert gewiß."*

[6] *"Das Erscheinen eines bestimmten Ortes x_0 bei der Beobachtung ... wird dann als außerhalb der Naturgesetze stehende Schöpfung aufgefaßt."*

between temporal and logical aspects of actual and 'counterfactual' measurements can be found in Mermin (1998).

The prime intention of this book is to discuss the relations between various arrows of time, and to search for a universal *master arrow*. To this end, certain open problems which have often been pragmatically put aside in the traditional theories will have to be clearly worked out. They may indeed become essential in more general theories, or have important cosmological implications (see Chaps. 5 and 6).

1

The Physical Concept of Time

The concept of time has been discussed since the earliest records of philosophy, when science had not yet become a separate subject. It is rooted in the subjective experience of the 'passing' present or moment of awareness, which appears to 'flow' through time and thereby to dynamically separate the past from the future. This has led to the formal representation of time by the real numbers, and to the picture of a present as a point that 'moves' in the direction defined by their sign.

The *mechanistic concept of time* is also based on this representation of time by the real numbers, but it avoids any subjective foundation: it is defined in terms of objective motion (in particular that of the celestial bodies). This concept is often attributed to Aristotle, although he seems to have regarded such a definition as insufficient.[1] A concept of time *defined* (not merely measured) by motion may indeed appear as a circular construction, since motion

---

[1] "Time is neither identical with movement nor capable of being separated from it" (Physics, Book IV). This may sound like an argument for some absoluteness of time. However, the traditional philosophical debate about time is usually linked to (and often confused with) the psychological and epistemological problem of the *awareness* of time 'in the soul', and hence related to the problem of consciousness. This is understandable, since ancient philosophers could not have anticipated the role of physico-chemical processes (that is, *motions*) in the brain as 'controlling the mind', and they were not in possession of reasonable clocks to give time a precise operational meaning for fast phenomena. According to Flasch (1993), Albertus Magnus (ca. 1200–1280) was the first philosopher who supported a rigorously 'physical' concept of time, since he insisted that time exists in Nature, while the soul merely perceives it: "Ergo esse temporis non dependet ab anima, sed temporis perceptio."

Another confusing issue of time in early philosophy, reflected by some of Zeno's paradoxes, was the mathematical problem of the real numbers, required to characterize the continuum. Before the discovery of calculus, mathematical concepts ('instruments of the mind') were often thought to be restricted to the natural numbers, while reality would correspond to the conceptually inaccessible continuum. Therefore, periodic motion was essential for *counting* time in order to grasp

is defined as change with (that is, dependence on) time, thus rendering the metaphor of the *flow of time* a tautology (see, e.g., Williams 1951). However, it forms a convenient *tool for comparing* different motions, provided an appropriate concept of simultaneous events is available. In pre-relativistic physics, this could be operationally defined by their simultaneous observation – later corrected for the time required for the propagation of light in a presumed 'ether'. (In German, an instant is called an *Augenblick*.) The possibility of comparing different motions, including clocks, indeed provides a sufficient basis for all meaningful temporal statements. All 'properties of time' must then be abstractions from *relative* motions and their empirical laws.

Physicists concerned with the concept of time have usually been quite careful in avoiding any hidden regress to the powerful prejudice of *absolute time*. Newton postulated it as a means to formulate his empirically founded laws, which then in turn justified this concept. More recent conceptions of time in physics may instead be understood as a *complete elimination* of absolute time, and hence of absolute motion. This approach is equivalent to the construction of 'timeless orbits', such as $r(\phi)$ for motion in a plane, which may be derived by eliminating t from the time-dependent solutions $r(t)$ and $\phi(t)$ of Newton's equations. In a similar way, all motions $q_i(t)$ in the Universe can be replaced by 'timeless' trajectories $q_i(q_0)$ in a global configuration space, where the hand of an appropriate 'clock' may be used as q_0.

These timeless trajectories may also be described by means of a *physically meaningless* parameter λ in the form $q_i(\lambda)$ for all i, where equal values of λ characterize the simultaneity of different q_i's. Such a parametric form was used by Jacobi to formulate his variational principle of mechanics (see Sect. 5.4), since astronomers without precise terrestrial clocks had to define time operationally as *ephemeris time* in terms of celestial motions obtained from their combined efforts (perturbation theory). If Jacobi's principle is applied to Newton's theory, absolute time can be recovered as a specific parameter λ that *simplifies* the equations of motion (Poincaré 1902). The existence of such a preferred time parameter, and its uniqueness up to linear transformations, is thus a non-trivial empirical property of Newtonian dynamics. It may then also be used to define equal time intervals at *different* times (as done by means of all conventional clocks, which *measure* this preferred time).

According to the most radical position about 'relational time', even its topology (ordering) has to be regarded as no more than the *consequence of this choice of an appropriate time parameter*. The 'timeless history' of the whole Universe would then be equivalent to an unordered 'heap of states' (or a stack of shuffled movie frames) that *can* be uniquely ordered and given a

---

it, not only to provide a measure. *Uniform* circular motion then appears as a natural assumption.

Since Newton, and even more so since Einstein, the concept of *time in Nature* has almost exclusively been elaborated by physicists. The adjective 'physical' in the title of this chapter is thus not meant as a restriction.

measure of distance only by the relations between their *intrinsic* structures (Barbour 1986, 1994a, 1999). This view will lead to entirely novel aspects in quantum gravity (see Sect. 6.2). If certain states from the stack (called 'time capsules' by Barbour) contain intrinsically consistent correlations representing memories, they may give rise to the impression of a flow of time to intrinsic observers, since the latter would remember properties of those global states which they interpret as forming their subjective past.

The concept of absolute motion thus shares the fate of the flow of time. 'Time reversal' is meaningful only as a *relative* reversal of motion (for example, relative to those physiological processes which control the subjective awareness of time and memory). Anyone who regards this mechanistic concept of time as insufficient should be able to explain what a reversal of *all* motion would *mean*. Ancient versions of a concept of time based on motion may have been understood as a 'causal control' of all motion on earth by the motions of (or on) the celestial spheres – an idea of which astrology is still a relic.

According to *Mach's principle* (see Barbour and Pfister 1995), the concept of absolute time is not only *kinematically* redundant – it should not even play any dynamical role as a preferred parameter, as it does in Newton's theory.[2] Similarly 'relativistic' ideas (although retaining an absolute concept of simultaneity) had already been entertained by Leibniz, Huygens, and Berkeley. They may even have prevented Leibniz from co-discovering Newton's mechanics, but led him to a definition of time in terms of *all* motions in the Universe. In this sense, an exactly periodic universe would describe the recurrence of the *same time*. This concept is far more rigorous than its ancient predecessor in not ascribing any preferred role to the motion of the celestial bodies.

Newton's mechanistic time, as used in his dynamical laws, specifies neither a direction in time nor a specific present. One may define a phenomenological direction by taking into account thermodynamical effects (including friction), thus arriving at the concept of a *thermodynamico-mechanistic time*. This concept is then based on the evidence that the thermodynamical arrow of time always and everywhere points in the same direction. Explaining this fact (or possibly its range of validity) must be part of the *physics of time asymmetry*. As will be explained, it can be understood within physics and cosmology, whereas physics does not even offer any conceptual means for deriving the concept of a present that would objectively separate the past from the future (see also the Epilog).

The concept of a present thus seems to have as little to do with the concept of time itself as color has to do with light (or with the nature of objects

---

[2] Mach himself was not very clear about whether he intended to postulate what is now often called his principle, or whether he intended to prove such a principle meaningless (see Norton 1995). A related confusion between the trivial invariance of a theory under a mere rewriting of the laws in terms of new spacetime coordinates ('Kretschmann invariance') and the nontrivial *invariance of the laws* under such coordinate transformations led to some dispute in early general relativity (Norton 1989).

reflecting it). Both the present and color characterize our subjective *perception* of time and light, respectively. Just as most information that is contained in the frequency spectrum of light being observed is lost in the eye or visual cortex before it may cause any brain activities associated with consciousness, all information about observed events which are separated in time by perhaps as much as two or three seconds seems to be combined to form certain neuronal 'states of being conscious' (see Pöppel, Schill, and von Steinbüchel 1990). The *moments of awareness* might thus even be discrete rather than reflecting the time continuum in terms of which the corresponding physical brain activities are successfully described. The time continuum remains a heuristic fiction – just like *all* concepts describing 'reality'. Similarly, the topology of colors (forming a closed circle), or the perception of different frequency mixtures of light as representing one and the same color, may readily be understood by means of physiological structures (see Goldsmith 2006, for example). However, neither the subjective appearance of colors (such as 'blue') nor that of the present can be derived from physical and physiological concepts. This non-trivial relationship between reality and the observed phenomena seems to assume an even more important and quite novel role in quantum descriptions – see Sects. 4.3 and 6.2.2. In contrast, the *direction* of the apparent 'passage' of time seems to be a consequence of the objective (thermodynamical) arrow that must also control neurobiological processes, and thus allows memories *of the past* to affect those 'states of being conscious'.

In Einstein's *special* theory of relativity, the mechanistic or thermo-dynamico-mechanistic concept of time may still be applied locally, that is, along time-like world lines. These *proper times*, although anholonomous (that is, path-dependent – as exemplified by the twin paradox), possess the hypothetical absoluteness of Newton's time, since they are assumed to be defined (or to 'exist') even in the absence of anything that may represent a clock. The claim of proper time as controlling all motion is formulated in the *principle of relativity*. While any *simultaneity* of spatially separate events represents no more than a choice of spacetime coordinates, local geometric and physical objects and properties can be defined 'absolutely'. An example is the *abstract* spacetime metric (to be distinguished from its basis-dependent representation by a matrix $g_{\mu\nu}$), which defines all proper times and the light cone structure. Hence, one may define a spacetime future and past *relative* to every spacetime point P (see Fig. 1.1), and unambiguously compare their orientations at different spacetime points by means of the path-independent parallel transport in this flat spacetime. So one may distinguish globally between past and future directions, and thus once again introduce a thermodynamico-mechanistic concept of time.[3]

---

[3] While superluminal objects ('tachyons') may be compatible with the relativistic light cone structure, they would pose severe problems to thermodynamics or the formulation of a physically reasonable boundary value problem (see Sect. 2.1).

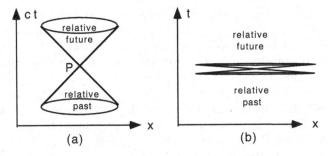

Fig. 1.1. (a) Local spacetime structure according to the theory of relativity. Space-time future and past are defined *relative* to every event P, and independent of any choice of reference frame. (b) In conventional units (large numerical value of the speed of light) the light cone opens widely, so its exterior seems to degenerate into a space-like hypersurface of 'absolute' simultaneity. What we *observe* as an apparently global present is in fact the backward light cone with respect to the subjective *here-and-now* P. Since only non-relativistic speeds are relevant in our macroscopic neighborhood, this apparent simultaneity then seems also to coincide with the forward light cone, that is, the spacetime border to the 'open' future that we (now) may affect by our 'free will' (things we can 'kick')

These consequences remain valid in *general* relativity if one excludes non-orientable manifolds, which would permit the continuous transport of forward light cones into backward ones. On the other hand, world lines may begin or end on spacetime singularities at finite values of their proper times. This prevents the applicability of Zermelo's recurrence objection that was raised against a statistical interpretation of thermodynamics (see Chap. 3). One may also have to avoid solutions of the Einstein equations which contain closed time-like curves (world lines which return into their own past *without* thereby changing their orientation). While compatible with general relativity, and even with flat spacetime if non-trivial topologies were considered, they would be incompatible with the usual assumption that the global past and future of an event exclude one another.

If local states of matter (such as described by fields) are unique functions on spacetime, a closed time-like curve must lead back to the *same* local state (including all memories and clocks). This would be inconsistent with a persisting thermodynamical arrow and/or 'free will' along closed world lines, and thus eliminate the much discussed murderer of his own grandfather when the latter was a child. Spacetime 'travel' is a misconception and a misleading picture that may require an external *second* concept of time – similar to the picture of a flowing time. Nonetheless, scenarios that would allow time travel are apparently quite popular even among professional relativists who do not care about thermodynamics. A 'spacetime traveler' would either have to stay forever on a loop in an exactly periodic manner (hence forming an exactly isolated reversible system), or to meet his older self already at his *first* arrival at their meeting point in spacetime. This would give rise to severe consistency problems if *all* irreversible phenomena (such as the documentation represented by retarded light) were consistently taken into account – in contrast to the usual science fiction stories. It is, therefore, not surprising that spacetime geometries with closed time-like curves seem to be *dynamically unstable* (and thus could never *arise*) in the presence of thermodynamically normal matter (Penrose 1969, Friedman et al. 1990, Hawking 1992, Maeda, Ishibashi, and Narita 1998). Closed time-like curves seem to be excluded by the same initial condition that is responsible for the arrow(s) of time. Other relations between thermodynamics and spacetime structure will be presented in Chap. 5.

If closed time-like curves are in fact excluded, then our spacetime can be time-ordered by means of a monotonic foliation. While there have been speculations about 'time warps' in *quantum gravity* (see Morris, Thorne and Yurtsever 1988, Frolov and Novikov 1990), their consistent description would have to take into account the rigorous revision of the concept of time that is a consequence of this theory (Sect. 6.2). A *quasi-classical* spacetime would have to presume the time arrow of decoherence for its justification (see Sects. 4.3 and 6.2.2). In quantum theory, the dynamically evolving state must be strongly entangled, that is, nonlocal (Sect. 4.2). There is then nothing to evolve locally (along time-like *curves* in spacetime).

The most important *novel* aspect of general relativity for the concept of time is the *dynamical* role played by spacetime geometry. It puts the geometry of space-like hypersurfaces in the position of 'physical objects' that evolve dynamically and interact with matter (see Sect. 5.4). In this way, spatial geometry itself becomes a *physical* clock, and the program of Leibniz and Mach may finally be fully taken into account by completely eliminating any relic of absolute time. While proper times (defined by means of the abstract metric) are traditionally regarded as a prerequisite for the formulation of dynamical laws, they are now *consequences of an evolving object* (the metric). In general relativity with matter, the spatial metric does not remain the *exclusive* definer of time as a controller of motion – although geometry still dominates over matter because of the large value of the Planck mass (see Sect. 6.2.2). This is reminiscent of Leibniz's elimination of the special role played by the celestial bodies, when he defined time in terms of *all* motion in the Universe.

This *physicalization of time* in accordance with Mach's principle (that may formally appear as its *elimination*) allows us even to speak of a direction *of* time instead of a direction *in* time – provided the spacetime of our Universe is clearly asymmetric. The dynamical role of geometry then also permits (and *requires*) the quantization of time (Sect. 6.2). Consequently, even the concept of a history of the Universe as a parametrizable succession of global states has to be abandoned. The conventional concept of time can at best be derived as a quasi-classical approximation.

General Literature: Reichenbach 1956, Mittelstaedt 1976, Whitrow 1980, Denbigh 1981, Barbour 1989, 1999.

The Time Arrow of Radiation

After a stone has been dropped into a pond, one observes concentrically diverging ('defocusing') waves. Similarly, *after* an electric current has been switched on, one finds a retarded electromagnetic field that is coherently propagating *away* from its source. Since the fundamental laws of Nature, which describe these phenomena, are invariant under time reversal, they are equally compatible with the reverse phenomena, in which concentrically focusing waves (and whatever may be dynamically related to them – such as heat) would 'conspire' in order to eject a stone out of the water. Deviations of the deterministic laws from time reversal symmetry would modify this argument only in detail (see the Introduction). However, the reversed phenomena are never observed in Nature. In high-dimensional configuration space, the absence of dynamical correlations which would focus to create local effects characterizes the time arrow of thermodynamics (Chap. 3), or, when applied to wave functions, even that of quantum theory (see Sect. 4.3).

Electromagnetic radiation will here be considered to exemplify wave phenomena in general. It may be described in terms of the four-potential A^μ, which in the Lorenz gauge obeys the wave equation

$$-\partial^\nu \partial_\nu A^\mu(\boldsymbol{r}, t) = 4\pi j^\mu(\boldsymbol{r}, t), \quad \text{with} \quad \partial^\nu \partial_\nu = -\partial_t^2 + \Delta, \qquad (2.1)$$

using units with $c = 1$, the notations $\partial_\mu := \partial/\partial x^\mu$ and $\partial^\mu := g^{\mu\nu}\partial_\nu$, and Einstein's convention of summing over identical upper and lower indices. When an appropriate boundary condition is imposed, one may write A^μ as a functional of the sources j^μ. For two well known boundary conditions one obtains the *retarded* and the *advanced* potentials,

$$A_{\text{ret}}^\mu(\boldsymbol{r}, t) = \int \frac{j^\mu(\boldsymbol{r}, t - |\boldsymbol{r} - \boldsymbol{r}'|)}{|\boldsymbol{r} - \boldsymbol{r}'|} \mathrm{d}^3 r', \qquad (2.2a)$$

$$A_{\text{adv}}^\mu(\boldsymbol{r}, t) = \int \frac{j^\mu(\boldsymbol{r}, t + |\boldsymbol{r} - \boldsymbol{r}'|)}{|\boldsymbol{r} - \boldsymbol{r}'|} \mathrm{d}^3 r'. \qquad (2.2b)$$

These two functionals of $j^\mu(\boldsymbol{r}, t)$ are related to one another by a reversal of retardation time $|\boldsymbol{r} - \boldsymbol{r}'|$ – see also (2.5) and footnote 4 below. Their linear combinations are again solutions of the wave equation (2.1).

At this point, many textbooks argue somewhat mysteriously that 'for reasons of causality', or 'for physical reasons', only the retarded fields, derived from the potential (2.2a) according to $F_{\text{ret}}^{\mu\nu} := \partial^\mu A_{\text{ret}}^\nu - \partial^\nu A_{\text{ret}}^\mu$, occur in Nature. This condition has therefore to be *added* to deterministic laws such as (2.1), which historically did indeed emerge from the asymmetric concept of causality. This example allows us to formulate in a preliminary way what seems to be meant by this *intuitive notion of causality*: correlated effects (that is, *nonlocal* regularities, such as coherent waves) must always possess a *local* common cause in their past.[1] However, this asymmetric notion of causality is a major *explanandum* of the physics of time asymmetry. As pointed out in the Introduction, it cannot be derived from the deterministic laws by themselves.

The popular argument that advanced fields are not found in Nature because they would require improbable initial correlations is known from statistical mechanics, but totally insufficient (see Chap. 3). The observed retarded phenomena are precisely as improbable among *all possible* ones, since they describe equally improbable *final* correlations. So their 'causal' explanation from an initial condition would beg the essential question.

Some authors take the view that retarded waves describe emission, advanced ones absorption. However, this claim ignores the fact that, for example, moving absorbers give rise to *retarded* shadows, that is, retarded waves which interfere destructively with incoming ones. In spite of the retardation, energy may thus flow from the electromagnetic field into an antenna. When incoming fields are present (as is generically the case), retardation does not necessarily mean emission of energy (see Sect. 2.1).

At the beginning of the last century, Ritz – following simular ideas by Planck and others – formulated a radical solution of the problem by postulating the exclusive existence of retarded waves *as a law*. Such time-directed *action at a distance* is equivalent to fixing the boundary conditions for the

---

[1] In the case of a *finite* number of local effects resulting from *one* local cause in the past, this situation is often viewed as a 'fork' in spacetime (see Horwich 1987, Sect. 4.8). However, this *fork of causality* should not be confused with the *fork of indeterminism* (in configuration space and time), which points to different (in general global) *potential states* rather than to different events (see also footnote 7 of Chap. 3 and Fig. 3.8). The fork of causality ('intuitive causality') may also characterize deterministic measurements and the documentation of their results, that is, the formation and distribution of information. It is related to Reichenbach's (1956) concept of *branch systems*, and to Price's (1996) *principle of independence of incoming influences* (PI[3]). Insofar as it describes the cloning and spreading of information, it represents an *overdetermination* of the past (Lewis 1986), or the *consistency of documents*. It is these correlations which let the macroscopic past appear 'fixed', while complete documents about *microscopic* history would be in conflict with thermodynamics and quantum theory.

electromagnetic field in a universal manner. The field would then *not* describe any degrees of freedom on its own, but just describe retarded forces.

This proposal, a natural generalization of Newton's gravitational force, led to a famous controversy with Einstein, who favored the point of view that retardation of radiation can be explained by thermodynamical arguments. Einstein, too, argued here in terms of an action-at-a-distance theory (see Sect. 2.4). At the end of their dispute, the two authors published a short letter in order to state their different opinions. After an introductory sentence, according to which retarded and advanced fields are equivalent "in some situations", the letter reads as what appears to be also a verbal compromise (Einstein and Ritz 1909 – my translation):[2]

> While Einstein believes that one may restrict oneself to this case without essentially restricting the generality of the consideration, Ritz regards this restriction as not allowed in principle. If one accepts the latter point of view, experience requires one to regard the representation by means of the retarded potentials as the only possible one, provided one is inclined to assume that the fact of the irreversibility of radiation processes has to be present in the laws of Nature. Ritz considers the restriction to the form of the retarded potentials as one of the roots of the Second Law, while Einstein believes that the irreversibility is exclusively based on reasons of probability.

Ritz thus conjectured that the thermodynamical arrow of time might be explained by the retardation of electromagnetic forces because of the latter's universal importance for all matter. However, the retardation of hydrodynamical waves (such as sound) would then have to be explained quite differently – for example, by again referring to the thermodynamical time arrow.

A similar but less well known controversy had already occurred in the nineteenth century between Max Planck and Ludwig Boltzmann. The former, at that time still an opponent of statistical mechanics, understood radiation as a genuine irreversible process, while the latter maintained that the problem is not different from that in kinetic gas theory: a matter of improbable initial conditions (Boltzmann 1897). These different interpretations became relevant, in particular, in connection with the quantum hypothesis: are quanta *caused*

---

[2] The original text reads: "*Während Einstein glaubt, daß man sich auf diesen Fall beschränken könne, ohne die Allgemeinheit der Betrachtung wesentlich zu beschränken, betrachtet Ritz diese Beschränkung als eine prinzipiell nicht erlaubte. Stellt man sich auf diesen Standpunkt, so nötigt die Erfahrung dazu, die Darstellung mit Hilfe der retardierten Potentiale als die einzig mögliche zu betrachten, falls man der Ansicht zuneigt, daß die Tatsache der Nichtumkehrbarkeit der Strahlungsvorgänge bereits in den Grundgesetzen ihren Ausdruck zu finden habe. Ritz betrachtet die Einschränkung auf die Form der retardierten Potentiale als eine der Wurzeln des Zweiten Hauptsatzes, während Einstein glaubt, daß die Nichtumkehrbarkeit ausschließlich auf Wahrscheinlichkeitsgründen beruhe.*"

by the emission process (as Planck had believed – later called quantum jumps – see Sects. 4.3.6 and 4.5), or inherent to light itself?

In Maxwell's classical *field theory*, the problem does not appear as obvious as in action-at-a-distance theories, since every bounded region of spacetime may contain 'free fields', which possess neither past nor future sources *in this region*. Therefore, one can consistently understand Ritz's hypothesis only cosmologically: *all* fields must possess advanced sources ('causes') somewhere in the Universe. While the examples discussed above demonstrate that the time arrow of radiation cannot merely reflect *the way* boundary conditions are posed, the problem becomes even more pronounced with the time-reversed question: "Do all fields also possess a *retarded source* (a *sink* in time-directed terms) somewhere in the future Universe?" This assumption corresponds to the *absorber theory of radiation*, a *T*-symmetric action-at-a-distance theory to be discussed in Sect. 2.4. The observed asymmetries would then require an unusual cosmic time asymmetry in the distribution of such sources.

2.1 Retarded and Advanced Form of the Boundary Value Problem

In order to distinguish the indicated pseudo-problem that concerns only the definition of 'free' fields from the physically meaningful question, one has to investigate the general boundary value problem for hyperbolic differential equations (such as the wave equation). This can be done by means of Green's functions, defined as the solutions of the specific inhomogeneous wave equation with a point-like source:

$$-\partial^\nu \partial_\nu G(r,t;r',t') = 4\pi\delta^3(r-r')\delta(t-t') \, , \tag{2.3}$$

and an appropriate boundary condition in space and time. Some of the concepts and methods to be developed below will be applicable in a similar form in Sect. 3.2 to the Liouville equations (Hamilton's equations applied to ensembles of states of mechanical systems). Using (2.3), a solution of the general inhomogeneous wave equation (2.1) may then be written as a functional of its sources:

$$A^\mu(r,t) = \int G(r,t;r',t')j^\mu(r',t') \, d^3r' \, dt' \, , \tag{2.4}$$

where the boundary condition for $G(r,t;r',t')$ determines that for $A^\mu(r,t)$, too. Retarded or advanced solutions are obtained from Green's functions G_{ret} and G_{adv}, which are given by

$$G_{\genfrac{}{}{0pt}{}{\mathrm{ret}}{\mathrm{adv}}}(r,t;r',t') := \frac{\delta(t-t'\pm|r-r'|)}{|r-r'|} \, . \tag{2.5}$$

The potentials A^μ_{ret} and A^μ_{adv} resulting from (2.4) are thus functionals of sources only on the past *or* future light cones of their argument, respectively.

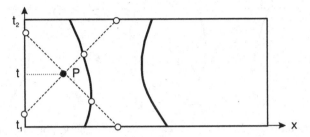

Fig. 2.1. Kirchhoff's boundary value problem, including initial, final and spatial boundaries. Sources (*thick world lines*) within the considered region and boundaries on both light cones (*dashed lines*) may in general contribute to the electromagnetic potential A^μ at the spacetime point P

By contrast, Kirchhoff's formulation of the boundary value problem allows one to express every specific solution $A^\mu(\boldsymbol{r}, t)$ of the wave equation by means of any Green's function $G(\boldsymbol{r}, t; \boldsymbol{r}', t')$. This can be achieved by using the three-dimensional Green theorem

$$\int_V \left[G(\boldsymbol{r}, t; \boldsymbol{r}', t')\Delta' A^\mu(\boldsymbol{r}', t') - A^\mu(\boldsymbol{r}', t')\Delta' G(\boldsymbol{r}, t; \boldsymbol{r}', t') \right] \mathrm{d}^3 r' \tag{2.6}$$

$$= \int_{\partial V} \left[G(\boldsymbol{r}, t; \boldsymbol{r}', t')\nabla' A^\mu(\boldsymbol{r}', t') - A^\mu(\boldsymbol{r}', t')\nabla' G(\boldsymbol{r}, t; \boldsymbol{r}', t') \right] \cdot \mathrm{d}\boldsymbol{S}' \,,$$

where $\Delta = \nabla^2$ is the Laplace operator, and ∂V is the boundary of the spatial volume V. Multiplying (2.3) by $A^\mu(\boldsymbol{r}', t')$, and integrating over $\boldsymbol{r}'$ and t' from t_1 to t_2 – on the right-hand side (RHS) by means of the δ-functions, while using the Green theorem and twice integrating by parts with respect to t' on the left-hand side (LHS), one obtains by further using (2.1):

$$A^\mu(\boldsymbol{r}, t) = \int_{t_1}^{t_2} \int_V G(\boldsymbol{r}, t; \boldsymbol{r}', t') j^\mu(\boldsymbol{r}', t') \, \mathrm{d}^3 r' \, \mathrm{d}t'$$

$$- \frac{1}{4\pi} \int_V \left[G(\boldsymbol{r}, t; \boldsymbol{r}', t')\partial_{t'} A^\mu(\boldsymbol{r}', t') - A^\mu(\boldsymbol{r}', t')\partial_{t'} G(\boldsymbol{r}, t; \boldsymbol{r}', t') \right] \mathrm{d}^3 r' \bigg|_{t_1}^{t_2}$$

$$+ \frac{1}{4\pi} \int_{t_1}^{t_2} \int_{\partial V} \left[G(\boldsymbol{r}, t; \boldsymbol{r}', t')\nabla' A^\mu(\boldsymbol{r}', t') - A^\mu(\boldsymbol{r}', t')\nabla' G(\boldsymbol{r}, t; \boldsymbol{r}', t') \right] \cdot \mathrm{d}\boldsymbol{S}' \, \mathrm{d}t'$$

$$\equiv \text{'source term'} + \text{'boundary terms'}. \tag{2.7}$$

if the event P described by $\boldsymbol{r}$ and t lies within the spacetime boundaries. Here, both (past and future) light cones may contribute to the three terms occurring in (2.7), as indicated in Fig. 2.1.

The formal T-symmetry of this representation of the potential as a sum of a source term and boundary terms in the past *and* future can be broken by the choice of Green's functions. When using one of the two forms (2.5), the

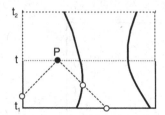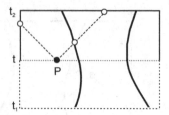

Fig. 2.2. Two representations of the same electromagnetic potential at time t by means of retarded or advanced Green's functions. They require data on partial boundaries (indicated by *solid lines*) corresponding to an initial or a final value problem, respectively

spacetime boundary required for determining the potential at time t assumes specific forms indicated in Fig. 2.2. Hence, the *same* potential can be written according to one or the other RHS of

$$A^\mu = \text{source term} + \text{boundary terms} = A^\mu_{\text{ret}} + A^\mu_{\text{in}}$$

$$= A^\mu_{\text{adv}} + A^\mu_{\text{out}} . \qquad (2.8)$$

For example, A^μ_{in} is here that solution of the *homogeneous* equations which coincides with A^μ for $t = t_1$. A^μ_{ret} and A^μ_{adv} vanish by definition for $t = t_1$ or $t = t_2$, respectively. *Any* field can therefore be described equivalently by an initial *or* a final value problem – with arbitrary boundary conditions. This result reflects the T-symmetry of the laws, while phenomenological causality is often used as an *ad hoc* argument for choosing G_{ret} rather than G_{adv}.

However, *two* free boundary conditions in the mixed form of Fig. 2.1 would in general not be consistent with one another, even if individually incomplete (see also Sects. 2.4 and 5.3). Retarded and advanced fields formally resulting from past and future sources, respectively, *do not add independently* (as sometimes assumed to describe a conjectured retro-causation) – they just contribute to different (or mixed) *representations* of the same field. In field theory, no (part of the) field 'belongs to' a certain source (in contrast to specific action-at-a-distance theories). Sources determine only the *difference* $A^\mu_{\text{out}} - A^\mu_{\text{in}}$ – similar to $T/i = S - 1$ in the interaction picture of the S-matrix. As can be seen from (2.8), this difference is identical to $A^\mu_{\text{ret}} - A^\mu_{\text{adv}}$. In causal *language*, where A^μ_{in} is regarded as given, the source 'creates' precisely its retarded field that has to be added to A^μ_{in} in the future of the source (where $A^\mu_{\text{adv}} = 0$).

Physically, *spatial* boundary conditions represent an interaction with the (often uncontrollable) spatial environment. For infinite spatial volume ($V = \mathbb{R}^3$), when the light cone cannot reach ∂V within finite time $t - t_1$, or in a closed universe, one loses this boundary term in (2.7), and thus obtains the *pure* initial value problem (for $t > t_1$),

$$A^\mu = A^\mu_{\text{ret}} + A^\mu_{\text{in}} \equiv \int_{t_1}^{t} \int_{\mathbb{R}^3} G_{\text{ret}}(\boldsymbol{r}, t; \boldsymbol{r}', t') j^\mu(\boldsymbol{r}', t') \, \mathrm{d}^3 r' \, \mathrm{d}t' \tag{2.9}$$

$$+ \frac{1}{4\pi} \int_{\mathbb{R}^3} \left[G_{\text{ret}}(\boldsymbol{r}, t; \boldsymbol{r}', t_1) \partial_{t_1} A^\mu(\boldsymbol{r}', t_1) - A^\mu(\boldsymbol{r}', t_1) \partial_{t_1} G_{\text{ret}}(\boldsymbol{r}, t; \boldsymbol{r}', t_1) \right] \mathrm{d}^3 r' ,$$

and correspondingly the pure final value problem $(t < t_2)$,

$$A^\mu = A^\mu_{\text{adv}} + A^\mu_{\text{out}} \equiv \int_{t}^{t_2} \int_{\mathbb{R}^3} G_{\text{adv}}(\boldsymbol{r}, t; \boldsymbol{r}', t') j^\mu(\boldsymbol{r}', t') \, \mathrm{d}^3 r' \, \mathrm{d}t' \tag{2.10}$$

$$- \frac{1}{4\pi} \int_{\mathbb{R}^3} \left[G_{\text{adv}}(\boldsymbol{r}, t; \boldsymbol{r}', t_2) \partial_{t_2} A^\mu(\boldsymbol{r}', t_2) - A^\mu(\boldsymbol{r}', t_2) \partial_{t_2} G_{\text{adv}}(\boldsymbol{r}, t; \boldsymbol{r}', t_2) \right] \mathrm{d}^3 r' .$$

The different signs at t_1 and t_2 are due to the fact that the gradient in the direction of the outward-pointing normal vector has now been written as a derivative with respect to t_1 (inward) or t_2 (outward).

So one finds precisely the retarded potential $A^\mu = A^\mu_{\text{ret}}$ if $A^\mu_{\text{in}} = 0$. (Only the 'Coulomb part', required by Gauß's law, must always be present by constraint. It can be regarded as the retarded *or* advanced consequence of the conserved charge.) In scattering theory, an initial condition fixing the incoming wave (usually described by a plane wave) is called a *Sommerfeld radiation condition*. Both conditions are to determine the actual situation. Therefore, the physical problem is not which of the two forms, (2.9) or (2.10), is *correct* (both are), but:

1. Why does the Sommerfeld radiation condition $A^\mu_{\text{in}} = 0$ (in contrast to $A^\mu_{\text{out}} = 0$) approximately apply in many situations?
2. Why are initial conditions more *useful* than final conditions?

The second question is related to the *historical nature* of the world. Answers to these questions will be discussed in Sect. 2.2.

The form (2.7) of the four-dimensional boundary value problem, characteristic of determinism in field theory, applies to partial differential equations of hyperbolic type (that is, with a Lorentzian signature $-+++$). Elliptic type equations would instead lead to the Dirichlet or von Neumann problems, which require values of the field *or* its normal derivative, respectively, on a *closed* boundary (which in spacetime would have to include past *and* future). Only hyperbolic equations lead generally to 'propagating' solutions, which are compatible with free initial conditions. They are thus responsible for the concept of a *dynamical state* of the field, which facilitates the familiar concept of time.

The wave equation (with its hyperbolic signature) is known to be derivable from Newton's equations as the continuum limit of a spatial lattice of mass points, held at their positions by means of harmonic forces. For a linear chain, $m \mathrm{d}^2 q_i / \mathrm{d}t^2 = -k \big[(q_i - q_{i-1} - a) - (q_{i+1} - q_i - a) \big]$ with $k > 0$, this is the limit $a \to 0$ for fixed ak and m/a. The crucial restriction to 'attractive' forces $(k > 0)$ may here appear surprising, since Newton's equations are *always* deterministic, and allow one to pose initial conditions regardless of the type or sign of the forces. However, only bound (here oscillating) systems possess a *stable* position (here characterized by the lattice constant a). In the same limit, an elliptic differential equation (with signature $++++$) would result for a lattice of variables q_i with repulsive forces ($k < 0$). This repulsion, though still representing deterministic dynamics, would cause the particle distances $q_i - q_{i-1}$ to explode immediately in the limit $k \to \infty$. The unstable solution $q_i - q_{i-1} = a$ is in this case the only eigensolution of the Dirichlet problem with eigenvalue 0 (derived from the *condition* of a bounded final state). Mathematically, the dynamically diverging solutions simply do not 'exist' any more in the continuum limit.

For second order wave equations, a hyperbolic signature forms the basis for all (exact or approximate) conservation laws, which give rise to the continuity of 'objects' in time (including the 'identity' of observers). For example, the free wave equation has solutions of a *conserved form* $f(z \pm ct)$, while the Klein–Gordon equation with a positive and variable 'squared mass' $m^2 = V(\mathbf{r}, t)$ has unitary solutions $i\partial\phi(\mathbf{r}, t)/\partial t = \pm\sqrt{-\Delta + V}\,\phi(\mathbf{r}, t)$. This dynamical consequence of the spacetime metric, which leads to such 'wave tubes' (see also Sect. 6.2.1), is crucial for what appears as the inevitable 'progression of time' (in contrast to our freedom to move in space). However, the *direction* of this apparent flow of time requires *additional* conditions.

This section was restricted to the boundary value problem for fields in the presence of *given* sources. In reality, the charged sources depend in turn on the fields by means of the Lorentz force. The resulting coupled system of differential equations is still T-symmetric, while all consequences of the retardation regarding the *actual* electromagnetic fields, derived in this and the following section, remain valid. New problems will arise, though, from the *self-interaction* of point charges or elementary charged rigid objects (see Sect. 2.3).

2.2 Thermodynamical and Cosmological Properties of Absorbers

Wheeler and Feynman (1945, 1949) took up the Einstein–Ritz controversy about the relation between the two time arrows of radiation and thermodynamics. Their work essentially confirms Einstein's point of view, provided his 'reasons of probability' are replaced by 'thermodynamical reasons'. Statistical reasons by themselves are insufficient for deriving a thermodynamical arrow (see Chap. 3.) The major part of Wheeler and Feynman's arguments were again based on a T-symmetric action-at-a-distance theory, which is particularly well suited for presenting them in an historical context. From the point of view of local field theory (that is for good reasons preferred today), this picture may appear strange or even misleading. The description of their *absorber theory of radiation* will therefore be postponed until Sect. 2.4.

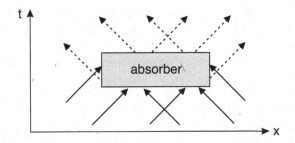

Fig. 2.3. Ideal absorbers do not contribute by means of G_{ret}. (*Arrows* represent the formal time direction of retardation)

In field theory, radiation is described by a continuum of variables, which may themselves require the application of thermodynamical concepts (as is well known for black body radiation). However, coupled harmonic oscillators are not ergodic, and so would not approach equilibrium. For this reason, radiation in a cavity consisting of reflecting walls was usually assumed to contain a small dust grain of coal in order to allow its spectral distribution to equilibrate by absorption and re-emission. I will here neglect the presence of reflecting bodies, and define absorbers (in the 'ideal' case assumed to possess infinite heat capacity) by the following phenomenological properties:

> A spacetime region is called an '(ideal) absorber' if any radiation propagating within its boundaries is (immediately) thermalized at the absorber temperature T ($= 0$).

The *thermalization* referred to in this definition is based on the arrow of time given by the Second Law. For electromagnetic waves this can also be described by means of a complex refractive index when using the Maxwell equations. The sign of its imaginary part reflects the thermodynamical arrow. The definition means that no radiation can propagate within ideal absorbers, and in particular that no radiation may leave the absorbing region along forward light cones. This consequence can then be applied to the boundary value problem as follows (see also Fig. 2.3):

> By means of the retarded Green's function, (ideal) absorbers forming parts of a spacetime boundary contribute only thermal radiation at the absorber temperature T ($= 0$).

Such a boundary condition simplifies the *initial* value problem considerably. If the space-like part ∂V of the boundary required for the retarded form of the boundary value problem depicted in Fig. 2.2 consists entirely of ideally absorbing walls (as is usually an excellent approximation for the relevant frequencies in a laboratory or other closed rooms), the condition $A^{\mu}_{\text{in}} = 0$ applies shortly after the initial time t_1 that is used to define the 'incoming' fields in (2.7). So one finds precisely the retarded fields (including reflected waves) of sources which are present in the laboratory. On the other hand, absorbers on the boundary would *not* affect contributions to the Kirchhoff problem by means of G_{adv}; in the nontrivial case one has $A_{\text{out}}^{\mu} \neq 0$. Therefore, in this laboratory situation the radiation arrow is a simple consequence of the thermodynamical arrow characterizing absorbers.

Do similar arguments also apply to situations outside absorbing boundaries, in particular in astronomy? The night sky does in fact appear black, representing a condition $A_{\text{in}}^{\mu} \approx 0$, although the present Universe is transparent to visible light. Can the darkness of the night sky then be understood in a realistic cosmological model? For the traditional model of an infinitely old universe this was impossible, a situation called *Olbers' paradox* after one of the first astronomers who mentioned this problem. The total brightness B of the sky beyond the atmosphere would then be given by

$$B = 4\pi \int_0^{\infty} \rho L_{\text{a}}(r) r^2 \, \mathrm{d}r \,, \tag{2.11}$$

where ρ is the number density of sources (mainly the fixed stars), while $L_{\text{a}}(r) = \bar{L}/r^2$ is their mean apparent luminosity. In the static and homogeneous situation ($\bar{L}, \rho = $ constant) this integral diverges linearly, and the night sky should be infinitely bright. Light absorption by stars in the foreground would reduce this result to a finite but large value, corresponding to a sky as bright as the mean surface of a star. It would not help to take into account other absorbing matter, since this would soon have to be in thermal equilibrium with the radiation under these conditions.

Olbers' paradox was resolved by Hubble's discovery of the expansion of the Universe, which required a finite age of the order of 10^{10} years (following a *big bang*). An integral of type (2.11) with a finite upper limit would in general remain bounded. Since all wavelengths λ grow proportional to the expansion parameter $a(t)$, this leads according to Wien's displacement law, $T \propto \lambda^{-1}$, to the reduction of the apparent temperature T_{a} of all past sources. Stefan and Boltzmann's law for thermal radiation, $L \propto T^4$, then requires that the apparent brightness of the stars, L_{a}, decreases not only with the geometric factor r^{-2}, but also with the inverse fourth power of a. In a homogeneous expanding universe of finite age, the brightness of the sky is then given by

$$B \propto \int_0^{\tau_{\text{max}}} \rho(t_0 - \tau) \bar{L}(t_0 - \tau) \left[\frac{a(t_0 - \tau)}{a(t_0)} \right]^4 \mathrm{d}\tau \,, \tag{2.12}$$

where t_0 means the present, while $\tau_{\text{max}} \approx t_0$ is the age of the transparent universe. If neither the total number of stars nor their mean absolute luminosity, $\bar{L}$, have changed, the integrand is simply proportional to $a(t_0 - \tau)/a(t_0)$. This or similar models lead to a negligible contribution from star light. Indeed, if our present universe were static, times of the order of 10^{23} yr, that is, exceeding the Hubble age by a factor of 10^{13}, would be required in the integral (2.11) to produce a night sky as bright as the surface of a mean star (Harrison 1977).

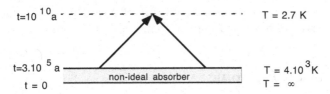

Fig. 2.4. The cosmological initial value problem for the electromagnetic radiation. The thermal contribution of the non-ideal absorber represented by the hot, ionized matter during the radiation era has now cooled down to the measured background radiation of 2.7 K (which can be neglected for most purposes)

While this conclusion resolves Olbers' original paradox, it is does *not* explain the cosmological condition $A^{\mu}_{\mathrm{in}} \approx 0$, since (1) it *presumes* retardation, and (2) the nature of sources must have drastically changed during the early history of the Universe. In its 'radiation era', matter was ionized and almost homogeneous, representing a non-ideal absorber with a temperature of several thousand degrees (see Fig. 2.4) that can serve as an initial boundary. Because of the cosmic expansion, the thermal radiation of this absorber has cooled down to its now observed value of 2.73 K, compatible with the darkness of the night sky.

The cosmic expansion, which is vital for this low present temperature, is thus also essential for the non-equilibrium formed by the contrast between cold interstellar space and the hot stars. The latter are producing their energy by nuclear reactions under the control of gravitational contraction – see Chap. 5. The expansion of the Universe has therefore often been proposed as the *master arrow* of time. However, it would be inappropriate to use *causal* arguments to explain this connection. Even in a presumed contraction era of the Universe, absorbers would then retain their intrinsic arrow of time. In order to reverse it, the thermodynamical arrow would have to be reversed, too. The scenario of fields and phenomenological absorbers in an expanding universe is far too simple to describe a master arrow. This cosmological discussion will therefore be resumed in Sects. 5.3 and 6.2.

In a quite different approach, Hogarth (1962) had suggested that the opacity of intergalactic matter (cosmic absorbers) must have *changed* drastically during the evolution of the Universe in order to provide a time asymmetry that would explain the observed retardation of radiation. Inspired by Wheeler and Feynman's time-symmetric definition of absorbers (Sect. 2.4), he neglected the thermodynamical arrow of absorbers. However, even in thermal equilibrium, a time arrow may survive in the form of correlations between microscopic variables unless *enforced* otherwise (see the Appendix for an example).

The above conclusions regarding the retardation of electromagnetic radiation apply accordingly to all kinds of waves in interaction with matter obeying thermodynamics. Only gravitational waves might be sufficiently decoupled from absorbers, since even the radiation era must have been transparent to them. Ritz's conjecture of a *law-like* nature of retarded electrodynamics will therefore be reconsidered and applied to gravity[3] in Chap. 5.

2.3 Radiation Damping

This somewhat technical section describes an important application of retarded fields. Except for Dirac's radiation reaction of (2.22), which will be used in Sect. 2.4, its results are rarely needed for the rest of the book.

The emission of electromagnetic radiation by a charged particle that is accelerated by an external force requires the particle to react by *losing* energy. Similar to friction, this *radiation reaction*, described by an *effective equation of motion*, must change sign under time reversal. As will be explained, this can be understood as a consequence of the retardation of the field when acting on its own source, even though the retardation seems to disappear at the position of a point source. However, the self-interaction of point-like charges leads to singularities (infinite mass renormalization) which need care when being separated from that part of the interaction which is responsible for radiation damping. While these problems could be avoided if any self-interaction were eliminated by means of the action-at-a-distance theory (described in Sect. 2.4), others would arise in their place.

Consider the trajectory of a charged particle, represented by means of its Lorentzian coordinates $z^\mu(\tau)$ as functions of proper time τ. The corresponding four-velocity and four-acceleration are $v^\mu := dz^\mu/d\tau$ and $a^\mu := d^2z^\mu/d\tau^2$, respectively.[4] From $v^\mu v_\mu = -1$ one obtains by differentiation $v^\mu a_\mu = 0$ and $v^\mu \dot{a}_\mu = -a^\mu a_\mu$. In a rest frame, defined by $v^k = 0$ (with $k = 1, 2, 3$), one has $a^0 = 0$.

The four-current density of this point charge is given by

$$j^\mu(x^\nu) = e \int v^\mu(\tau)\delta^4[x^\nu - z^\nu(\tau)]d\tau . \tag{2.13}$$

Its retarded field $F_{ret}^{\mu\nu} = 2\partial^{[\mu}A_{ret}^{\nu]} := \partial^\mu A_{ret}^\nu - \partial^\nu A_{ret}^\mu$ is known as the Liénard–Wiechert field. The retarded or advanced fields can be written in an invariant

---

[3] The retardation of gravitational waves has been indirectly confirmed by double pulsars (see Taylor 1994).

[4] While an orbit in space or configuration space would merely be passed backwards under time reversal ($t \rightarrow -t$), a worldline in spacetime *changes* according to $z^k(\tau) \rightarrow z^k(-\tau)$ and $z^0(\tau) \rightarrow -z^0(-\tau)$ (for $k = 1, 2, 3$). The reversal of the parameter τ is now only a consequence of the convention $dt/d\tau > 0$, but physically meaningless. As the derivative $v^\mu(\tau)$ – and accordingly also the current $j^\mu(\tau)$ – then get an additional minus sign under time reversal, the potentials A^μ and fields $F^{\mu\nu}$ inherit this transformation property for their respective indices (corresponding to $\boldsymbol{E} \rightarrow \boldsymbol{E}$ and $\boldsymbol{B} \rightarrow -\boldsymbol{B}$). In order to study questions of (ir)reversibility, one may often better use the simpler TP transformations, $z^\mu(\tau) \rightarrow -z^\mu(-\tau)$ for all μ – see the Introduction.

manner (see, for example, Rohrlich 1965) as

$$F^{\mu\nu}_{\text{ret/adv}}(x^\sigma) = \pm \frac{2e}{\rho} \frac{\mathrm{d}}{\mathrm{d}\tau} \frac{v^{[\mu}R^{\nu]}}{\rho}$$

$$= \frac{2e}{\rho^2} v^{[\mu}u^{\nu]} + \frac{2e}{\rho} \left\{ a^{[\mu}v^{\nu]} - u^{[\mu}v^{\nu]}a_u \pm u^{[\mu}a^{\nu]} \right\} , \qquad (2.14)$$

with v^μ and a^μ taken at times τ_{ret} or τ_{adv}, respectively. In this expression,

$$R^\mu := x^\mu - z^\mu\big(\tau_{\text{ret/adv}}\big) =: (u^\mu \pm v^\mu)\rho , \qquad (2.15)$$

with $u^\mu v_\mu = 0$ and $u^\mu u_\mu = +1$, is the light-like vector pointing from the retarded or advanced spacetime position z^μ of the source to the point x^μ where the field is considered. Obviously, ρ is the distance in space or in time between these points in the rest frame of the source, while $a_u := a^\mu u_\mu$ is the component of the acceleration in the direction of the unit spatial distance vector u^μ. Retardation or advancement are enforced by the condition of R^μ being light-like, that is, $R^\mu R_\mu = 0$.

On the RHS of (2.14), second line, the field consists of two parts, proportional to $1/\rho^2$ and $1/\rho$. They are called the generalized Coulomb field ('near-field') and the radiation field ('far-field'), respectively. Since the stress–energy tensor

$$T^{\mu\nu} = \frac{1}{4\pi} \left(F^{\mu\alpha}F^\nu_\alpha + \frac{1}{4}g^{\mu\nu}F^{\alpha\beta}F_{\alpha\beta} \right) \qquad (2.16)$$

is quadratic in the fields, it then consists of *three* parts characterized by different powers of ρ. For example, one has

$$T^{\mu\nu}_{\text{ret}} := T^{\mu\nu}(F^{\mu\nu}_{\text{ret}}) = \frac{e^2}{4\pi\rho^4} \left(u^\mu u^\nu - v^\mu v^\nu - \frac{1}{2}g^{\mu\nu} \right)$$

$$+ \frac{e^2}{2\pi\rho^3} \left\{ a_u \frac{R^\mu R^\nu}{\rho^2} - \big[v^{(\mu}a_u + a^{(\mu}\big]\frac{R^{\nu)}}{\rho} \right\}$$

$$+ \frac{e^2}{4\pi\rho^2}(a_u^2 - a^\lambda a_\lambda)\frac{R^\mu R^\nu}{\rho^2} , \qquad (2.17)$$

where braces around pairs of indices define symmetrization, so for example, $v^{(\mu}R^{\nu)} := (v^\mu R^\nu + v^\nu R^\mu)/2$. Here, $T^{\mu\nu}$ is the ν-component of the current of the μ-component of four-momentum. In particular, T^{0k} is the Poynting vector in the chosen Lorentz system, and $T^{\mu\nu}\mathrm{d}^3\sigma_\nu$ is the flux of four-momentum through an element $\mathrm{d}^3\sigma_\nu$ of a hypersurface. If $\mathrm{d}^3\sigma_\nu$ is space-like (a volume element), this 'flux' describes its energy–momentum ('momenergy') content, otherwise it is the flux through a spatial surface element during an element of time.

The retarded field caused by an element of the world line of the point charge between τ and $\tau + \Delta\tau$ has its support between the forward light cones of these two points, that is, on a thin four-dimensional conic shell (see Fig. 2.5).

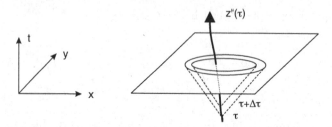

Fig. 2.5. The spacetime support of the retarded field of a world line element $\Delta\tau$ of a point charge is located between two light cones (co-axial only in the rest frame of the source). The flux of field momentum crosses light cones in the near-field region of the charge

The intersection of the cones with a space-like hyperplane forms a spherical shell (concentric only in the rest frame at time τ, and in the figure depicted two-dimensionally as a narrow ring). The integral of the stress–energy tensor over this spherical spatial shell,

$$\Delta P^\mu = \int T^{\mu\nu}\mathrm{d}^3\sigma_\nu \,, \tag{2.18}$$

is the four-momentum of the field on this hyperplane 'caused' by the world line element Δz^μ. In general, this momentum is not conserved along light cones, since (2.17) contains a momentum flux orthogonal to the cones, due to the dragging of the near-field by the charge. Therefore, Teitelboim 1970 suggested a time-asymmetric splitting of the energy–momentum tensor, which leads to an asymmetric electron dressing – valid *only* in connection with given F_{in}. However, the flux component orthogonal to the cones vanishes in the far-zone, where $T^{\mu\nu}$ is proportional to $R^\mu R^\nu$. In this region the integral (2.18) describes the four-momentum radiated *away* from the trajectory of the charge during the interval $\Delta\tau$,

$$\Delta P^\mu \xrightarrow[\rho\to\infty]{} \Delta P^\mu_{\mathrm{rad}} = \frac{2}{3}e^2 a^\lambda a_\lambda v^\mu \Delta\tau =: \Re v^\mu \Delta\tau \,. \tag{2.19}$$

The quantity $\Re = 2e^2 a^\lambda a_\lambda/3$ is called the *invariant rate of radiation*. In the comoving rest frame ($v^k = 0$), one recovers the non-relativistic Larmor formula,

$$\Delta P^0_{\mathrm{rad}} = v_\mu \Delta P^\mu_{\mathrm{rad}} = \frac{2}{3}e^2 a^\lambda a_\lambda \Delta t = \frac{2}{3}e^2 \mathbf{a}^2 \Delta t \,. \tag{2.20}$$

This result confirms that the energy transfer into radiation in a positive interval of time cannot be negative – a consequence of the presumed retardation. An accelerated charged particle must *lose* energy to radiation, regardless of the direction of the driving *external* force.

Larmor's formula led to a certain confusion when it was applied to a charged particle in a gravitational field. Because of its dependence on acceleration, (2.19) is restricted to inertial frames. In general relativity, inertial frames are freely falling ones. According to the principle of equivalence, a freely falling charge should then *not* radiate, while a charge 'at rest' in a gravitational field (under the influence of non-gravitational forces) should do so. This problem was not understood until Mould (1964) demonstrated that the response of a detector to radiation depends on its acceleration, too (see also Fugmann and Kretzschmar 1991).

In general relativity, the principle of equivalence is only locally valid (see Rohrlich 1963). However, a *homogeneous* gravitational field (as would result from a homogeneous massive plane) is described by a flat spacetime, and thus globally equivalent to a *rigid field of uniform accelerations* a^μ on Minkowski spacetime. This field corresponds to a set of 'parallel' hyperbolic trajectories with constant (in time, but varying between trajectories) accelerations $a^\mu a_\mu$. These trajectories define accelerated rigid frames, since they preserve distances in comoving frames. Together with their proper times, the trajectories define the curved *Rindler coordinates* – see (5.16) and Fig. 5.5 in Sect. 5.2.

The equivalence principle can therefore be *globally* applied to a homogeneous gravitational field. This means that an inertial (freely falling) detector is *not* excited by an inertial charge, while a detector 'at rest' is. The latter would remain idle in the presence of a charge being 'equivalently at rest' (at a fixed distance in this case). A detector-independent definition of *total radiation* also turns out to depend on acceleration (as it should for consistency) because of the occurrence of spacetime horizons for truly uniform acceleration (see Boulware 1980 and Sect. 5.2).

The emission of energy according to (2.20) thus requires a deceleration of the point charge in order to conserve total energy. It should be possible to derive this consequence directly from the fundamental dynamical equations, which are governed by the Lorentz force,

$$\mathcal{F}^\mu_{\mathrm{self}}(\tau) = eF^{\mu\nu}_{\mathrm{ret}}\big[z^\sigma(\tau)\big]v_\nu(\tau)\,, \tag{2.21}$$

resulting from the particle's self-field. However, this expression leads to problems caused by the fact that the electromagnetic force acts only on the point charge, where the self-field is singular (its Coulomb part even with $1/\rho^2$), while part of the accelerated mass is contained in the energy of the comoving Coulomb field. Paul Dirac (1938) showed that the symmetric part $\bar{F}^{\mu\nu}$ of the retarded field,

$$F^{\mu\nu}_{\mathrm{ret}} = \frac{1}{2}(F^{\mu\nu}_{\mathrm{ret}} + F^{\mu\nu}_{\mathrm{adv}}) + \frac{1}{2}(F^{\mu\nu}_{\mathrm{ret}} - F^{\mu\nu}_{\mathrm{adv}}) \equiv: \bar{F}^{\mu\nu} + F^{\mu\nu}_{\mathrm{rad}}\,, \tag{2.22}$$

is responsible for the infinite mass renormalization, while the antisymmetric part, $F^{\mu\nu}_{\mathrm{rad}}$, remains regular, and indeed describes the radiation reaction when treated properly.

In order to prove the second part of this statement, one has to expand all quantities in (2.14) up to the third order in terms of the retardation $\Delta\tau_{\mathrm{ret}} = \tau_{\mathrm{ret}} - \tau$, e.g.,

$$v^\nu(\tau_{\text{ret}}) = v^\nu(\tau + \Delta\tau_{\text{ret}}) = v^\nu(\tau) + \Delta\tau_{\text{ret}}a^\mu(\tau) \tag{2.23}$$

$$+\frac{1}{2}\Delta\tau_{\text{ret}}^2\dot{a}^\mu(\tau) + \frac{1}{6}\Delta\tau_{\text{ret}}^3\ddot{a}^\mu(\tau) + \cdots .$$

All terms which are singular at the position of the point charge cancel from the antisymmetric field, and one obtains (see Rohrlich 1965, p. 142)

$$F_{\text{rad}}^{\mu\nu} = -\frac{4e}{3}\dot{a}^{[\mu}v^{\nu]} . \tag{2.24}$$

The resulting PT-antisymmetric Lorentz self-force, the *Abraham four-vector*

$$\mathcal{F}_{\text{rad}}^\mu := eF_{\text{rad}}^{\mu\nu}v_\nu = \frac{2e^2}{3}(\dot{a}^\mu + v^\mu\dot{a}^\nu v_\nu) = \frac{2e^2}{3}(\dot{a}^\mu - v^\mu a^\nu a_\nu) \tag{2.25}$$

(using $a^\nu v_\nu = 0$ in the second step), should then describe the *radiation reaction* of a point charge. It leads to a nonlinear equation of motion (the *Lorentz–Abraham–Dirac or LAD equation*). However, while the second term on the RHS of (2.25) is in accord with (2.19), the $\dot{a}^\mu$ term is ill-defined (see below). Together with the singular mass renormalization term resulting from $\bar{F}^{\mu\nu}$, it describes the four-momentum transfer from the point charge itself to its comoving singular near-field.

In a rest frame (with $v^k = 0$ and $a^0 = 0$), one obtains

$$\mathcal{F}_{\text{rad}}^0 = -\frac{2e^2}{3}a^2 , \qquad \mathcal{F}_{\text{rad}}^k = \frac{2e^2}{3}\frac{da^k}{dt} . \tag{2.26}$$

Therefore, the radiation reaction describes non-relativistically a force proportional to the change of acceleration, da^k/dt, while its fourth component is the energy *loss* according to the non-negative invariant rate of radiation (2.19). The latter was originally defined by the energy flux through a distant sphere on the future light cone (Fig. 2.5). However, global conservation laws may be used only if all their contributions are taken into account. For example, one would not obtain an analogous conservation of *three-momentum* for the bare point charge and its far-field because of the aforementioned momentum flux orthogonal to the future light cone of the moving charge. For this reason, the uniformly accelerated charge may radiate with $\Re \neq 0$ even though the 'radiation reaction' $\mathcal{F}_{\text{rad}}^\mu$ (including its ill-defined term) vanishes in this case, as can be seen separately for its two non-vanishing components, $\mathcal{F}_{\text{rad}}^\mu a_\mu$ and $\mathcal{F}_{\text{rad}}^\mu v_\mu$.

If the boundary condition $F_{\text{in}}^{\mu\nu} = 0$ does not hold, the complete electromagnetic force acting on a point charge is given by

$$ma^\mu = \mathcal{F}^\mu = \mathcal{F}_{\text{in}}^\mu + \mathcal{F}_{\text{rad}}^\mu = \mathcal{F}_{\text{out}}^\mu - \mathcal{F}_{\text{rad}}^\mu \tag{2.27}$$

– cf. Sect. 2.1 and (2.22). Terms caused by the symmetric part of the self-field have now been brought to the LHS in the form of a mass renormalization $\Delta m a^\mu$. Equation (2.27) still exhibits T-symmetry, but the latter may be broken fact-like by the given *initial* condition $F_{\text{in}}^{\mu\nu}$ (in contrast to the uncontrollable outgoing radiation contained in $F_{\text{out}}^{\mu\nu}$). The LAD equation, based on (2.25) and the *first* RHS of (2.27), may then be written in the form

$$m(a^\mu - \tau_0 \dot{a}^\mu) = K^\mu(\tau) := \mathcal{F}^\mu_{\text{in}} - \Re v^\mu \,, \tag{2.28}$$

where $\tau_0 = 2e^2/3mc^2$ is the time required for light to travel a distance of the order of the 'classical electron radius' e^2/mc.

Both terms of (2.28) that result from the radiation reaction (2.25) now change sign under time reversal (or the interchange of retarded and advanced fields). While the second one (now on the RHS) is the friction-type radiation damping $-\Re v^k$, required for the conservation of energy, the one now appearing on the LHS (called the *Schott term*) is proportional to the *third* time-derivative of the position in an inertial frame. A solution to the LAD equation (2.28) would thus require *three* initial vectors as integration constants (the initial acceleration in addition to the usual initial position and velocity). Evidently, information has been lost by differentiation in the expansion (2.23). Even for $\mathcal{F}^\mu_{\text{in}} = 0$, the LAD equation (2.28) admits *runaway* solutions, non-relativistically in the form of an exponentially increasing *self-acceleration*, $a^k(t) = a^k(0) \exp(t/\tau_0)$.

Because of this formal information loss, the LAD equation is *not* a complete equation of motion. It can only represent a necessary condition for the motion of the point charge. In the free case, unphysical runaway solutions could simply be eliminated by fixing the artificial integration constant by the condition $a^k(0) = 0$. However, this would still lead to runaway as soon as an external force were turned on, since the formal solution of (2.28) with respect to a^μ is

$$ma^\mu(\tau) = e^{\tau/\tau_0}\left[ma^\mu(0) - \frac{1}{\tau_0}\int_0^\tau e^{-\tau'/\tau_0} K^\mu(\tau')d\tau'\right]. \tag{2.29}$$

Therefore, Dirac suggested fixing the initial acceleration in terms of the *future force* according to $ma^\mu(0) = (1/\tau_0)\int_0^\infty e^{-\tau'/\tau_0} K^\mu(\tau')d\tau'$. The substitution $\tau' \to \tau' + \tau$ then leads to Dirac's equation of motion,

$$ma^\mu(\tau) = \int_0^\infty K^\mu(\tau + \tau')\frac{e^{-\tau'/\tau_0}}{\tau_0}d\tau'. \tag{2.30}$$

It represents a Newtonian (second order) equation of motion which depends on a force that acts ahead of time. How could this 'acausal' result be derived using retarded fields alone?

Moniz and Sharp (1977) demonstrated that the pathological behavior of this 'classical electron' is a consequence of a mass renormalization that exceeds the physical electron mass (so that the bare mass must be negative). If the point charge is replaced by a rigid charged sphere of radius r_0 in its rest frame, one obtains, by using the now everywhere regular retarded field, an equation of motion that was first proposed by Caldirola (1956), and later derived by Yaghjian (1992) as an approximation. It reads

$$m_0 a^\mu(\tau) = \mathcal{F}^\mu_{\text{in}}(\tau) + \frac{2e^2}{3r_0}\frac{v^\mu(\tau - 2r_0) + v^\mu(\tau)v^\nu(\tau)v_\nu(\tau - 2r_0)}{2r_0}, \tag{2.31}$$

where m_0 is the bare mass. The retardation $2r_0$ in the arguments would change sign for advanced fields (consistent only in conjunction with given $\mathcal{F}_{\text{out}}^\mu$). Taylor expansion of (2.31) with respect to $2r_0$, equivalent to (2.23), and using $v^\mu v_\mu = -1$ and its time derivatives leads in first order to a finite mass renormalization (4/3 of the electrostatic mass), and in second order back to the LAD equation (see Zeh 1999a). While (2.31) is analogous to a non-Markovian master equation (see Sect. 3.2), the LAD equation corresponds to its Markovian limit, valid for slowly varying fields. In this sense, the radiation reaction has to be calculated from the given history in order to *determine* the acceleration (rather than its derivative) towards the future (right derivative).

The self-force acting on the rigid 'electron' according to (2.31) is the difference (because of $v^\mu v_\mu = -1$) between a decelerating and an accelerating friction type force with different retardations. For positive bare and physical masses it does not lead to runaway, although it may possess complicated non-analytic solutions, in particular for forces varying on a time scale shorter than the light travel time within the charged sphere. Dirac's pre-acceleration of the center of mass can now be understood as a consequence of the *presumed* rigidity of the charged sphere, which requires forces of constraint acting ahead of time.

The most rigorous elimination of unphysical solutions from the LAD equation so far was proposed by Spohn (2000) – see also Rohrlich (2001), while the history of electron theory is discussed in Rohrlich (1997). It seems that the concept of a non-inertial point charge is inconsistent with classical electrodynamics, while external forces acting on a charge *distribution* would disturb its shape and structure. A quantum ground state of the electron may instead be protected against deformations by its discrete excitation spectrum. However, an explicit QED eigenstate would have to include nonlocal *quantum entanglement* between particle and field modes in an essential way (see Sect. 4.2).

General Literature: Rohrlich 1965, 1997, Levine, Moniz and Sharp 1977, Boulware 1980.

2.4 The Absorber Theory of Radiation

Ritz's retarded action-at-a-distance theory, mentioned at the beginning of this chapter, eliminates all electromagnetic degrees of freedom by postulating the *cosmological* initial condition $F_{\text{in}}^{\mu\nu} = 0$ in order to fix all forces of electromagnetic origin. Since electromagnetic forces would then act only on the forward light cones of their sources, this theory cannot be compatible with Newton's third law, which requires their reactions. However, the reaction to a retarded action must be advanced.[5] In order to warrant energy–momentum conserva-

---

[5] In *field theory*, sources and fields interact *locally* in spacetime. For this reason the self-force (2.25) could not be derived from the flux of field momentum in the far-zone.

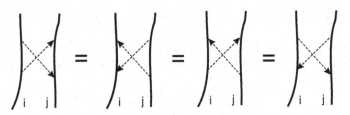

Fig. 2.6. Different interpretations of the same interaction term of the Hamiltonian for a pair of particles

tion, an action-at-a-distance theory has to be formulated in a T-symmetric way, as done by Fokker (1929) by means of his action

$$I = \int (T - V)\mathrm{d}t = \sum_i m_i \int \mathrm{d}\tau_i \tag{2.32}$$

$$-\frac{1}{2} \sum_{i \neq j} e_i e_j \iint v_i^\mu v_{j\mu} \delta\big[(z_i^\nu - z_j^\nu)(z_{i\nu} - z_{j\nu})\big] \mathrm{d}\tau_i \mathrm{d}\tau_j \; .$$

Here, indices i and j are particle numbers. A sum over $i \neq j$ defines a double sum excluding equal indices, while a sum over $i(\neq j)$ is meant as a sum over i only, excluding a given value j. In (2.32), the particle positions z_i^μ and velocities v_i^μ have to be taken at the proper time τ_i of the corresponding particle, for example $z_i^\mu = z_i^\mu(\tau_i)$.

Expanding the δ-function in the potential energy according to

$$\delta(\Delta z^\nu \Delta z_\nu) = \delta(\Delta z_0^2 - \Delta \mathbf{z}^2) = \frac{1}{2|\Delta \mathbf{z}|}\Big[\delta(\Delta z_0 - |\Delta \mathbf{z}|) + \delta(\Delta z_0 + |\Delta \mathbf{z}|)\Big] \tag{2.33}$$

(with $\Delta z^\nu = z_i^\nu - z_j^\nu$) preserves its symmetric form. By integrating either over τ_i or over τ_j, one obtains, respectively, the first or second of the following expressions (first two graphs of Fig. 2.6):

$$\frac{e_i}{2} \int \Big[A_{\mathrm{ret},j}^\mu(z_i^\sigma) + A_{\mathrm{adv},j}^\mu(z_i^\sigma)\Big] v_{i\mu} \mathrm{d}\tau_i \equiv \frac{e_j}{2} \int \Big[A_{\mathrm{adv},i}^\mu(z_j^\sigma) + A_{\mathrm{ret},i}^\mu(z_j^\sigma)\Big] v_{j\mu} \mathrm{d}\tau_j \; . \tag{2.34}$$

$A_{\mathrm{ret},j}^\mu$ and $A_{\mathrm{adv},j}^\mu$ are the retarded and advanced potentials of the jth particle according to (2.2a) and (2.2b). However, if the integral is always carried out with respect to the particle on the backward light cone of the other one, one obtains, in spite of the preserved T-symmetry of the theory, only contributions in terms of retarded potentials (third graph):

$$\frac{e_i}{2} \int A_{\mathrm{ret},j}^\mu(z_i^\sigma) v_{i\mu} \mathrm{d}\tau_i + \frac{e_j}{2} \int A_{\mathrm{ret},i}^\mu(z_j^\sigma) v_{j\mu} \mathrm{d}\tau_j \; , \tag{2.35}$$

and analogously, but time reversed, for the advanced potentials (fourth graph). Einstein seems to have been referring to this equivalence of different *forms* of the interaction in his letter with Ritz (quoted in the introduction to this chapter).

However, the Euler–Lagrange equations resulting from (2.32) automatically lead to T-symmetric forces which comply with Newton's third law:

$$ma_i^\mu = \frac{e_i}{2} \sum_{j(\neq i)} \left[F_{\text{ret},j}^{\mu\nu}(z_i^\sigma) + F_{\text{adv},j}^{\mu\nu}(z_i^\sigma) \right] v_{i,\nu} \ . \tag{2.36}$$

According to (2.8), this would correspond to the cosmic boundary condition $F_{\text{in}}^{\mu\nu} + F_{\text{out}}^{\mu\nu} = 0$ in Maxwell's theory. Equations (2.36) differ from the empirically required ones,

$$ma_i^\mu = e_i \sum_{j(\neq i)} F_{\text{ret},j}^{\mu\nu}(z_i^\sigma) v_{i,\nu} + \frac{e_i}{2} \left[F_{\text{ret},i}^{\mu\nu}(z_i^\sigma) - F_{\text{adv},i}^{\mu\nu}(z_i^\sigma) \right] v_{i,\nu} \ , \tag{2.37}$$

not only by the replacement of half the retarded by half the advanced forces, but also by the missing radiation reaction $\mathcal{F}_{\text{rad},i}^\mu$ (Dirac's asymmetric self-force). While the problem of a mass renormalization has disappeared, (2.36) seems to be in drastic conflict with reality. Moreover, it contains a complicated dynamical meshing of the future with the past that does not in any obvious way permit the formulation of an initial-value problem.

The two equations of motion, (2.36) and (2.37), differ precisely by a force that would result from the sum of the asymmetric fields of *all* particles, $F_{\text{rad,total}}^{\mu\nu} = \sum_j (F_{\text{ret},j}^{\mu\nu} - F_{\text{adv},j}^{\mu\nu})/2$. Since the retarded and advanced fields appearing in this expression possess identical sources, their difference solves the *homogeneous* Maxwell equations, and thus represents a *free* field in spite of the dependence of the retarded and advanced fields on the sources. Therefore, this sum of differences may be assumed to vanish for *all* times as a 'boundary' condition. As there are no retarded fields at the beginning of the Universe, this would require $\sum_j F_{\text{adv},j}^{\mu\nu}(t_{\text{big bang}}) = 0$ as a very restrictive *global constraint on all sources* that will ever arise; it can hardly be exactly valid.

If the condition $F_{\text{rad,total}}^{\mu\nu} = 0$ did apply, the advanced effects of all charged matter in the Universe would precisely double the retarded forces in (2.36), cancel the advanced ones, and imitate a self-interaction that is responsible for radiation damping. This is an example of the equivalence of apparently quite different dynamical representations of *deterministic* theories, such as causal or teleological, local or global ones.

Instead of referring to a cosmic initial condition, Wheeler and Feynman (1945) tried to explain the vanishing of the sum of asymmetric fields by the assumption that the total charged matter in the Universe behaves as an 'absorber' in a sense that is very different from that used in Sect. 2.2. They required that the symmetric field $\bar{F}$ resulting from *all* particles, which would according to (2.36) determine the force on an additional 'test particle', should vanish *for statistical reasons* (by destructive interference) in a presumed empty space surrounding all matter of this 'island universe'. This assumption,

$$\sum_j \bar{F}_j^{\mu\nu} := \sum_j \frac{1}{2}\left[F_{\text{ret},j}^{\mu\nu} + F_{\text{adv},j}^{\mu\nu}\right] \longrightarrow 0 \, , \tag{2.38}$$

constitutes their cosmic *absorber condition*. Since the retarded or advanced fields vanish by definition in the asymptotic past or future, respectively, so must their time-reversed partner because of (2.38), and hence also their asymmetric combination. Wheeler and Feynman then concluded by means of the homogeneous Maxwell equations that the total asymmetric field would vanish *everywhere*. This is just the required 'boundary' condition.

However, the consistency of this procedure is very questionable. A similar problem would arise for an expanding and recollapsing Universe that were sandwiched between two *thermodynamically opposite* radiation eras (absorbers with opposite thermodynamical arrows of time) – see Sect. 5.3. As explained in Sect. 2.1, the compatibility of double-ended (two-time) boundary conditions is highly nontrivial – similar to an eigenvalue problem. This consistency problem is particularly severe for a universe that remains optically transparent and thus preserves information contained in the radiation such as light (Davies and Twamley 1993).

In contrast to the physical absorbers of Sect. 2.2, the new *absorber condition* is symmetric under time reversal. This fact led to many misunderstandings. For example, rather than adding the vanishing antisymmetric term to (2.36), one might as well subtract it in order to obtain the time-reversed representation

$$ma_i^\mu = e_i \sum_{j(\neq i)} F_{\text{adv},j}^{\mu\nu}(z_i^\sigma)v_{i,\nu} - \frac{e_i}{2}\left[F_{\text{ret},i}^{\mu\nu}(z_i^\sigma) - F_{\text{adv},i}^{\mu\nu}(z_i^\sigma)\right]v_{i,\nu} \, . \tag{2.39}$$

Although it is as correct as (2.37) under the absorber condition, (2.39) describes *advanced* actions and a radiation reaction that leads to reverse damping (exponential acceleration).

Therefore, Wheeler and Feynman's absorber condition cannot explain the observed radiation arrow. Neither (2.37) nor (2.39) would describe the local empirical situation, which requires in general that only a limited number of 'obvious sources' contribute noticeably to the *retarded* sum (2.37). Otherwise, retardation would never have been recognized. This means that the retarded contribution of all 'other' sources (those which form the true universal absorber) must interfere destructively (see Fig. 2.7):

$$\sum_{i \in \text{absorbers}} F_{\text{ret},i}^{\mu\nu} \approx 0 \qquad \text{'inside' universal absorber} \, . \tag{2.40}$$

This is possible (except for the remaining thermal radiation) if the absorber particles approach thermal equilibrium by means of collisions *after* having been accelerated by retarded fields. Therefore, one *cannot* expect

$$\sum_{i \in \text{absorbers}} F_{\text{adv},i}^{\mu\nu} \approx 0 \qquad \text{'inside' universal absorber} \tag{2.41}$$

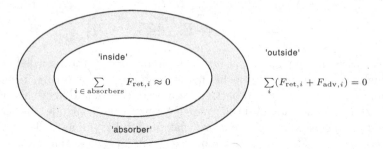

Fig. 2.7. T-symmetric ('outside') and T-asymmetric ('inside') absorber conditions of a model Universe with action-at-a-distance electrodynamics

to hold in a symmetric way. Since $F_{\text{ret}}^{\mu\nu}$ contributes only on the forward light cones, Fig. 2.7 reduces to Fig. 2.4.

In order to justify the applicability of (2.37) in contrast to that of (2.39), one still needs the *asymmetric* condition that has been derived in Sect. 2.2 from the thermodynamical arrow of time under certain cosmological assumptions. This means that any motion of absorber particles is dissipated as heat *after* it has been induced. While in field theory the field may be regarded as 'matter' with its own thermodynamical state, action-at-a-distance theory ascribes thermodynamical properties only to the sources. In the former description, the relation between electromagnetic and thermodynamical arrows is just an example of the *universality* of the thermodynamical arrow (see Sect. 3.1.2).

With these remarks I also hope to put to rest objections raised by Popper (1956) against the thermodynamical foundation of the radiation arrow – see also Price (1996), page 51. The only 'unusual' aspect of electromagnetic fields (when regarded as matter) is their weak coupling, which may greatly delay their thermalization in the absence of absorbers (see also Sect. 5.3.3). It is this very property that allows light and radio waves to serve as information media.

Therefore, the time-reversal-symmetric 'absorber condition' (2.38) leads to the equivalence of various forms of electrodynamics, but *cannot* explain the time arrow of radiation. In action-at-a-distance theories, there is no free radiation, while the radiation reaction is the effect of advanced forces 'caused' by future absorbers. If the Universe remained transparent for all times in some direction, an appropriately beamed emitter should not draw any power according to the absorber theory. As it always seems to do so (Partridge 1973), the absorber theory may even be ruled out empirically. If similarly applied to *gravitational* fields, it might also be in conflict with the observed energy loss of double pulsars.

General Literature: Wheeler and Feynman 1945, 1949, Hoyle and Narlikar 1995.

3

The Thermodynamical Arrow of Time

The thermodynamical arrow of time is characterized by the increase of entropy according to the Second Law. This law was first postulated by Rudolf Clausius in 1865 as a consequence of Carnot's theorem of 1824 when combined with the just established equivalence of heat with other forms of energy (the First Law of thermodynamics). It can be written in a general form by means of a sum of external and internal changes of entropy as

$$\frac{dS}{dt} = \left\{\frac{dS}{dt}\right\}_{\text{ext}} + \left\{\frac{dS}{dt}\right\}_{\text{int}} ,$$

where

$$dS_{\text{ext}} = \frac{dQ}{T} \quad \text{and} \quad \left\{\frac{dS}{dt}\right\}_{\text{int}} \geq 0 . \tag{3.1}$$

Here, S is phenomenologically defined as the entropy of a bounded system – thereby exploiting reversible processes with $(dS/dt)_{\text{int}} = 0$, while dQ is the reversible (infinitely slow) inward heat flux through the system's complete boundary during a time interval dt. (See also the local form (3.39) of the Second Law on p. 60.)

Conventionally, the heat flux is not written as a derivative dQ/dt, since its integral $Q(t)$ would not represent a 'function of state' – although it does, of course, define the time-integrated net flux in the actual process. The first term of dS/dt in (3.1) vanishes by definition for 'thermodynamically closed' systems. Since the whole Universe is defined as an absolutely closed system (even if infinite), its total entropy, or the mean entropy of co-expanding volume elements, should according to this law evolve towards its maximum – the so-called *Wärmetod* (heat death) of the world. The phenomenological thermodynamical concepts used in (3.1), in particular the temperature, apply only in situations of partial (local) equilibrium.

Statistical physics is now believed to provide an explanation and potential generalization of phenomenological thermodynamics – including its Second Law. While in principle *all* physical concepts are phenomenological, this term is used here to emphasize the presumed existence of conceivably complete microscopic concepts that await the application of statistical methods.

While statistical considerations are indeed essential for the understanding of thermodynamical concepts, statistics as a method of counting has nothing a priori to do with dynamics. Therefore, it cannot by itself explain dynamically 'irreversible' processes – characterized by $\{dS/dt\}_{int} > 0$. This requires *additional* assumptions, which often remain unnoticed, since they appear 'natural' to our prejudiced way of thinking in terms of 'causes' (exclusively in the past). These hidden assumptions have therefore to be carefully investigated in order to reveal the true origin of the thermodynamical arrow.

An attempt to explain this fundamental asymmetry on the basis of the 'historical nature' of the world, that is, by using the idea that the past is 'fixed' (and therefore neither requires nor allows statistical retrodiction) would clearly represent a circular argument when starting from nothing but time-symmetrically deterministic laws. This idea must itself be rooted in the time asymmetry of the physical world. The existence of reliable knowledge or information only about the past corresponds to a time-asymmetric *physical* relation between documents and their sources, analogous to the asymmetric 'causal' relation between retarded electromagnetic fields and charged currents as their advanced sources (Sect. 2.1). For example, light contains information about objects in the more or less recent past. Similarly, *all* documents represent an asymmetry in the physical world, and do not simply reflect the way boundary conditions (such as initial or final conditions) are posed.

In a statistical description, 'irreversible' processes are of the form

$$\text{improbable state} \xrightarrow{\ t\ } \text{probable state},$$

where the probability ratio is usually a huge number. These probabilities are defined by the size or measure of certain *sets* of elementary states (called 'representative ensembles' by Tolman 1938), which contain the *real* state of the considered system (a point in its microscopic configuration space) as a member. This measure of probability changes as the state moves along its trajectory through different such sets. If the representative ensembles are *operationally* defined, for example by means of macroscopic preparation procedures, they are often themselves called macroscopic or thermodynamical 'states'. This terminology has its origin in a description that is unaware of the microscopic states (or rejects the concept of such a microscopic reality). The dynamical justification of thermodynamical states is a major objective of a microscopic foundation of thermodynamics: why are certain sets of microscopic states 'representative' in forming macroscopic states?

Irreversible processes of the above kind would statistically be more abundant than those of the kind

$$\text{improbable state} \xrightarrow{\ t\ } \text{improbable state}.$$

Their overwhelming occurrence in Nature can therefore be understood under the presumption of improbable *initial* states. In an operational approach, such an assumption would simply be taken for granted: a consequence of operations to be performed in time. In a cosmological context it requires a cosmic initial condition. This has occasionally been called the *Kaltgeburt* (cold birth) of the Universe, although a low temperature (kT much smaller than energies of mechanical degrees of freedom) need not be its essential aspect – see Sect. 5.3. However, this initial assumption appears quite unreasonable precisely for statistical reasons, since (1) there are just as many processes of the type

$$\text{probable state} \xrightarrow[t]{} \text{improbable state},$$

and (2) far more of the kind

$$\text{probable state} \xrightarrow[t]{} \text{probable state}.$$

The latter describe equilibrium. Hence, for statistical reasons we should expect the world to *be* in the situation of a heat death, while the required improbable initial condition needs an explanation that does *not presume* causality.

The first of these two arguments is the 'reversibility objection' (*Umkehreinwand*), formulated by Boltzmann's friend and teacher Johann Joseph Loschmidt. It is based on the fact that each trajectory has precisely one time-reversed counterpart.[1] If, for example, $z(t) \equiv \{q_i(t), p_i(t)\}_{i=1,...,3N}$ describes a trajectory in $6N$-dimensional phase space (Γ-*space*) according to the Hamiltonian equations, then the time-reversed trajectory, $z_T(-t) \equiv \{q_i(-t), -p_i(-t)\}$, is also a solution of the equations of motion. If the entropy S of a state z can be defined as a function of this state, $S = F(z)$, with $F(z) = F(z_T)$, then Loschmidt's objection means that for every solution with $dS/dt > 0$ there is precisely one corresponding solution with $dS/dt < 0$. In statistical theories, $F(z)$ is defined as a monotonic function (conveniently the logarithm) of the size or measure of the mentioned set of states to which z belongs. The property $F(z) = F(z_T)$ is then a consequence of the symmetry character of the transformation $z \rightarrow z_T$, while the stronger objection (2) above means that there are far more solutions with $dS/dt \approx 0$, that is, $S(t) \approx S_{max}$ – simply because this condition characterizes almost all of configuration space.

In order to justify the thermodynamical arrow of time statistically, one therefore has to either derive the improbable initial conditions from an independent (time-asymmetric) cosmological assumption, or simply postulate them in some form. The Second Law is by no means *incompatible* with deterministic or T-symmetric dynamical laws; it is just extremely improbable and

---

[1] Often, T (or CPT) symmetry of the dynamics is assumed for this argument. This has misleadingly given rise to the by no means justified expectation that the difficulties in deriving the Second Law may be overcome by dropping this symmetry. However, as already pointed out in the Introduction, the crucial point in Loschmidt's argument is the time reversal symmetry of *determinism* itself (not of its precise form), which is often reflected by the possibility of compensating time reversal by another symmetry operation (see also Sect. 3.4).

in conflict with unbiased statistical reasoning. The widespread 'double standard' of readily accepting improbable initial conditions while rejecting similar final ones has been duly criticized by Price (1996).

Another argument against the statistical interpretation of irreversibility, the *recurrence objection* (or *Wiederkehreinwand*), was raised much later by Ernst Friedrich Zermelo, a collaborator of Max Planck at a time when the latter still opposed atomism, and instead supported the 'energeticists', who attempted to understand energy and entropy as fundamental 'substances'. This argument is based on a mathematical theorem due to Henri Poincaré, which states that every bounded mechanical system will return as close as one wishes to its initial state within a sufficiently large time. The entropy of a closed system would therefore have to return to its former value, provided only the function $F(z)$ is continuous. This is a special case of the *quasi-ergodic theorem* which asserts that every system will come arbitrarily close to any point on the hypersurface of fixed energy (and possibly with other fixed analytical constants of the motion) within finite time.

While all these theorems are mathematically correct, the recurrence objection fails to apply to reality for quantitative reasons. The age of our Universe is much smaller than the Poincaré recurrence times even for a gas consisting of no more than a few tens of particles. Their recurrence to the vicinity of their initial states (or their coming close to any other similarly specific state) can therefore be excluded in practice. Nonetheless, some 'foundations' of irreversible thermodynamics in the literature rely on formal idealizations that would lead to *strictly infinite* Poincaré recurrence times (for example the 'thermodynamical limit' of infinite particle number). Such assumptions are not required in our Universe of finite age, and they would *not* invalidate the reversibility objection (or the equilibrium expectation, mentioned above). However, all foundations of irreversible behavior have to presume some very improbable initial conditions.

The theory of thermodynamically irreversible processes must therefore address two main problems:

1. The investigation of realistic *mechanisms* which describe the dynamical evolution away from certain (presumed) improbable initial states. This is usually achieved in the form of 'master equations', which mimic a *law-like* T-asymmetry – analogous to Ritz's retarded action-at-a-distance in electrodynamics. In contrast to electrodynamics, they describe the dynamics of ensembles, equivalent to an effective stochastic dynamics for the individual states (applicable in the 'forward' direction of time). These mechanisms should be able to justify the *representative ensembles* (macroscopic states) and even describe the emergence of *order* (Sect. 3.4).

2. The precise nature of the required improbable initial states. This leads again to the quest for an appropriate *cosmic* initial condition, similar to the global condition $A_{in}^{\mu} = 0$ in the early Universe that would be able to explain the radiation arrow (see Sects. 2.2 and 5.3).

3.1 The Derivation of Classical Master Equations

Statistical physics is concerned with systems consisting of a large number of microscopic constituents which are known to obey quantum mechanics. However, quantum theory is still haunted by interpretational problems, in particular regarding the nature of probabilistic 'quantum events'. These are usually understood as representing a *fundamental* irreversible part of dynamics, that might even be the true source of thermodynamical irreversibility. In contrast, classical mechanics is deterministic and well defined. Therefore, *classical* statistical mechanics will be formulated and discussed in this chapter for conceptual consistency and later comparison with quantum statistical mechanics – even though it is based on an incorrect microscopic theory. Most thermodynamic properties of a gas, for example, can in fact be modelled by a system of interacting classical mass points – see (4.21). While the present section follows historical routes, a more general and systematic formalism, that can later also be used in quantum theory, will be presented in Sect. 3.2.

3.1.1 μ-Space Dynamics and Boltzmann's H-Theorem

The complete *dynamical state* of a mechanical system of N classical particles (distinguishable mass points) can either be represented by one point in its $6N$-dimensional phase space ('Γ-space'), or by N numbered points in six-dimensional 'μ-space' (the single-particle phase space). These N points form a *discrete distribution* in μ-space. If the particles are *not* distinguished from one another, this is exactly equivalent to an ensemble of $N!$ points in Γ-space that results from all particle permutations. Because of the large number of particles forming macroscopic systems (of order 10^{23}), Boltzmann (1866, 1896) used continuous (smoothed) distributions (or *phase space densities*) $\rho_\mu(\boldsymbol{p}, \boldsymbol{q})$ to describe them. This plausible approximation will turn out to have important consequences.

Two types of argument are in general used to justify it:

1. The formal *thermodynamical limit* $N \to \infty$. This represents an idealization that would lead to infinite Poincaré recurrence times. Mathematical proofs may then appear rigorous, while in fact they are approximations – valid only for the early (far from equilibrium) stage of our Universe. Though often convenient, this procedure may conceal physically important aspects, in particular when interchanging the thermodynamical limit with the limit $t \to \infty$ (physically a quantitative question).
2. Slightly 'uncertain' positions and momenta, defining small volume elements in Γ-space, $\Delta V_\Gamma = (\Delta V_\mu)^N$, instead of points. They describe infinite *ensembles* of states, and they may again lead to smooth distributions, since $N!$ such volume elements easily overlap even for a dilute gas, as $N!\Delta V_\Gamma \approx (N\Delta V_\mu)^N$ according to Stirling's approximation. Although uncertainties slightly larger than distances between the particles are sufficient for the smoothing, they will turn out to have drastic dynamical

consequences for many interacting particles. However, these uncertainties *cannot* be based on the quantum mechanical uncertainty relations with their corresponding phase space cells of size h^{3N}, since equivalent problems reappear in quantum theory if phase space points are consistently replaced by wave functions (see Sect. 4.1.1).

The time dependence of an individual point $\{p_i(t), q_i(t)\}$ in Γ-space (with $i = 1, \ldots, 3N$), described by Hamilton's equations, is equivalent to the simultaneous time dependence of all N points in μ-space. Therefore, the time dependence of an ensemble in Γ-space (represented by a distribution ρ_Γ) determines that of the corresponding density ρ_μ. In contrast to the dynamics in Γ-space (Sect. 3.1.2), however, this dynamics is not 'autonomous': the time derivative of a non-singular density ρ_μ is *not* determined by ρ_μ. The reason is that ρ_Γ cannot be recovered from ρ_μ in order to determine the latter's time derivative from that of the former. The mapping of Γ-space distributions on μ-space distributions cannot be uniquely inverted, as it destroys information about correlations between the particles (see also Fig. 3.1 and the subsequent discussion). The smooth μ-space distribution may, for example, characterize a 'macroscopic state' in the sense mentioned in the introduction to this chapter. Therefore, the envisioned chain of computation

$$\rho_\mu \longrightarrow \rho_\Gamma \xrightarrow{\ H\ } \frac{\mathrm{d}\rho_\Gamma}{\mathrm{d}t} \longrightarrow \frac{\partial \rho_\mu}{\partial t} \ , \tag{3.2}$$

which would be required to derive an autonomous dynamics for ρ_μ, is broken at its first link. Boltzmann's attempt to bridge this gap by statistical arguments will turn out to be the source of the time direction asymmetry in his statistical mechanics, and similarly in other formulations of irreversible processes. His procedure specifies a direction in time in a phenomenologically justified way, although it was originally meant to represent a general approximation rather than a modification of the Hamiltonian dynamics. One must then ask under what circumstances it may be valid.

Boltzmann *postulated* a stochastic dynamical law of the form

$$\frac{\partial \rho_\mu}{\partial t} = \left\{ \frac{\partial \rho_\mu}{\partial t} \right\}_{\text{free+ext}} + \left\{ \frac{\partial \rho_\mu}{\partial t} \right\}_{\text{collision}} . \tag{3.3}$$

Its first term is defined to describe particle motion under external forces only. It can be written as a continuity equation in 6-dimensional μ-space:

$$\left\{ \frac{\partial \rho_\mu}{\partial t} \right\}_{\text{free+ext}} = -\mathrm{div}_\mu j_\mu := -\nabla_q \cdot (\dot{q} \rho_\mu) - \nabla_p \cdot (\dot{p} \rho_\mu)$$

$$= -\nabla_q \cdot \left(\frac{p}{m} \rho_\mu \right) - \nabla_p \cdot (F_{\text{ext}} \rho_\mu) \ , \tag{3.4}$$

where j_μ is the current density in μ-space. In the absence of particle interactions this equation would describe the dynamics of the 'phase space fluid' exactly. It represents the *local* conservation of probability in μ-space according to the deterministic Hamiltonian equations, which hold separately for each particle in this case. Each point in μ-space (each single-particle state) moves continuously on a trajectory that is governed by the external forces $\boldsymbol{F}_{\text{ext}}$, thereby retaining its individual probability which was determined by the initial condition for ρ_μ.

For the second (non-trivial) term, Boltzmann proposed his *Stoßzahlansatz* (collision equation), which will be formulated here for simplicity under the following assumptions:

(1) $\boldsymbol{F}_{\text{ext}} = 0$ 'no external forces',

(2) $\rho_\mu(\boldsymbol{p}, \boldsymbol{q}, t) = \rho_\mu(\boldsymbol{p}, t)$ 'homogeneous distribution'.

The second condition is dynamically consistent for translation-invariant forces. From these assumptions one obtains $\{\partial\rho_\mu/\partial t\}_{\text{free+ext}} = 0$. The *Stoßzahlansatz* is then written in the plausible form

$$\frac{\partial\rho_\mu}{\partial t} = \left\{\frac{\partial\rho_\mu}{\partial t}\right\}_{\text{collision}} = \text{gains} - \text{losses}, \tag{3.5}$$

that is, as a *balance equation*. Its two terms on the RHS can be explicitly written in terms of transition rates $w(\boldsymbol{p}_1\boldsymbol{p}_2; \boldsymbol{p}'_1\boldsymbol{p}'_2)$ for particle pairs scattered from $\boldsymbol{p}'_1\boldsymbol{p}'_2$ to $\boldsymbol{p}_1\boldsymbol{p}_2$. They are usually (in a low density approximation) determined by the two-particle scattering cross-section, and they have to satisfy certain conservation laws. Because of this description in terms of rates for discontinuous changes of momenta, the collisions cannot be described by a *local* conservation of probability in μ-space, as in (3.4).

This *Stoßzahlansatz* (3.5) reads explicitly

$$\frac{\partial\rho_\mu(\boldsymbol{p}_1, t)}{\partial t} = \int \Big[w(\boldsymbol{p}_1\boldsymbol{p}_2; \boldsymbol{p}'_1\boldsymbol{p}'_2)\rho_\mu(\boldsymbol{p}'_1, t)\rho_\mu(\boldsymbol{p}'_2, t)$$

$$- w(\boldsymbol{p}'_1\boldsymbol{p}'_2; \boldsymbol{p}_1\boldsymbol{p}_2)\rho_\mu(\boldsymbol{p}_1, t)\rho_\mu(\boldsymbol{p}_2, t)\Big] \mathrm{d}^3 p_2 \mathrm{d}^3 p'_1 \mathrm{d}^3 p'_2 . \tag{3.6}$$

It forms the prototype of a *master equation* as an irreversible balance equation based on probabilistic transition rates. Because of their time asymmetry, these master equations cannot be generally valid approximations. They may hold for special solutions, which thus characterize an arrow of time. These solutions cannot even be particularly frequent among all other solutions.

For further simplification, invariance of the transition rates under *collision inversion*,

$$w(\boldsymbol{p}_1\boldsymbol{p}_2; \boldsymbol{p}'_1\boldsymbol{p}'_2) = w(\boldsymbol{p}'_1\boldsymbol{p}'_2; \boldsymbol{p}_1\boldsymbol{p}_2), \tag{3.7}$$

will be assumed. It may be derived from invariance under space reflection *and* time reversal, although these two symmetries do not necessarily have to be separately valid. The *Stoßzahlansatz* then assumes the form

$$\frac{\partial \rho_\mu(\boldsymbol{p}_1, t)}{\partial t} = \int w(\boldsymbol{p}_1 \boldsymbol{p}_2; \boldsymbol{p}_1' \boldsymbol{p}_2') \Big[\rho_\mu(\boldsymbol{p}_1', t) \rho_\mu(\boldsymbol{p}_2', t)$$

$$- \rho_\mu(\boldsymbol{p}_1, t) \rho_\mu(\boldsymbol{p}_2, t) \Big] \mathrm{d}^3 p_2 \mathrm{d}^3 p_1' \mathrm{d}^3 p_2' . \tag{3.8}$$

In order to demonstrate the irreversibility described by the *Stoßzahlansatz*, it is useful to consider *Boltzmann's H-functional*

$$H[\rho_\mu] := \int \rho_\mu(\boldsymbol{p}, \boldsymbol{q}, t) \ln \rho_\mu(\boldsymbol{p}, \boldsymbol{q}, t) \mathrm{d}^3 p \, \mathrm{d}^3 q = N \overline{\ln \rho_\mu} , \tag{3.9}$$

proportional to the mean logarithm of probability. The mean $\bar{f}$ of a function $f(\boldsymbol{p}, \boldsymbol{q})$ is defined here as $\bar{f} := \int f(\boldsymbol{p}, \boldsymbol{q}) \rho_\mu(\boldsymbol{p}, \boldsymbol{q}) \mathrm{d}^3 p \, \mathrm{d}^3 q / N$, in accordance with the normalization $\int \rho_\mu(\boldsymbol{p}, \boldsymbol{q}) \mathrm{d}^3 p \, \mathrm{d}^3 q = N$. Because of this fixed normalization, the H-functional is large for narrow distributions, but small for wide ones. An ensemble of discrete points (or δ-distributions), for example, would lead to $H[\rho_\mu] = \infty$, while a constant distribution on a region of volume V_μ, $\rho_\mu = N/V_\mu$, gives $H[\rho_\mu] = N(\ln N - \ln V_\mu)$. Note that H is defined only up to an additive constant that depends on the choice of a unit volume element of phase space in (3.10).

One may now derive Boltzmann's *H-theorem*,

$$\frac{\mathrm{d}H[\rho_\mu]}{\mathrm{d}t} \leq 0 , \tag{3.10}$$

by differentiating $H[\rho_\mu]$ with respect to time, while using the collision equation in the form (3.8):

$$\frac{\mathrm{d}H[\rho_\mu]}{\mathrm{d}t} = V \int \frac{\partial \rho_\mu(\boldsymbol{p}_1, t)}{\partial t} \big[\ln \rho_\mu(\boldsymbol{p}_1, t) + 1 \big] \mathrm{d}^3 p_1$$

$$= V \int w(\boldsymbol{p}_1 \boldsymbol{p}_2; \boldsymbol{p}_1' \boldsymbol{p}_2') \big[\rho_\mu(\boldsymbol{p}_1', t) \rho_\mu(\boldsymbol{p}_2', t) - \rho_\mu(\boldsymbol{p}_1, t) \rho_\mu(\boldsymbol{p}_2, t) \big]$$

$$\times \big[\ln \rho_\mu(\boldsymbol{p}_1, t) + 1 \big] \mathrm{d}^3 p_1 \mathrm{d}^3 p_2 \mathrm{d}^3 p_1' \mathrm{d}^3 p_2' . \tag{3.11}$$

The last expression may be conveniently reformulated by using the symmetries under collision inversion given by (3.7), and under particle permutation, $w(\boldsymbol{p}_1 \boldsymbol{p}_2; \boldsymbol{p}_1' \boldsymbol{p}_2') = w(\boldsymbol{p}_2 \boldsymbol{p}_1; \boldsymbol{p}_2' \boldsymbol{p}_1')$. (Otherwise this combined symmetry would be required to hold for short *chains* of collisions, at least.) Rewriting the integral in (3.11) as a symmetric sum of the four equivalent permutations of the integration variables, one obtains

$$\frac{\mathrm{d}H[\rho_\mu]}{\mathrm{d}t} = \frac{V}{4} \int w(\boldsymbol{p}_1 \boldsymbol{p}_2; \boldsymbol{p}_1' \boldsymbol{p}_2') \big[\rho_\mu(\boldsymbol{p}_1', t) \rho_\mu(\boldsymbol{p}_2', t) - \rho_\mu(\boldsymbol{p}_1, t) \rho_\mu(\boldsymbol{p}_2, t) \big]$$

$$\times \Big\{ \ln \big[\rho_\mu(\boldsymbol{p}_1, t) \rho_\mu(\boldsymbol{p}_2, t) \big] - \ln \big[\rho_\mu(\boldsymbol{p}_1', t) \rho_\mu(\boldsymbol{p}_2', t) \big] \Big\} \mathrm{d}^3 p_1 \mathrm{d}^3 p_2 \mathrm{d}^3 p_1' \mathrm{d}^3 p_2' \leq 0 .$$

$$\tag{3.12}$$

This integrand is manifestly non-positive, since the logarithm is a monotonically increasing function of its argument. This completes the proof of (3.10), which would apply to *any* monotonic function, not just the logarithm.

In order to recognize the relation between the H-functional and entropy, one may consider the *Maxwell distribution* ρ_M, given by

$$\rho_M(\boldsymbol{p}) := \frac{N}{V} \frac{\exp(-p^2/2mkT)}{\sqrt{(2\pi mkT)^3}} \ . \tag{3.13}$$

Its H-functional $H[\rho_M]$ has two important properties:

1. It represents a *minimum for given energy*, $E = \int \rho_\mu(\boldsymbol{p})[p^2/2m]\mathrm{d}^3p \approx \sum_i \boldsymbol{p}_i^2/2m$. A proof will be given in a somewhat more general form in Sect. 3.1.2. (Statistical reasoning unconstrained by a given energy value would predict infinite energy, since the phase space volume grows non-relativistically as its $(3N/2)$th power.) ρ_M must therefore represent an equilibrium distribution (with maximum entropy) under the *Stoßzahlansatz* if the transition probabilities are assumed to conserve energy.
2. One obtains explicitly

$$H[\rho_M] = V \int \rho_M(\boldsymbol{p}) \ln \rho_M(\boldsymbol{p}) \mathrm{d}^3p$$

$$= -N \left(\ln \frac{V}{N} + \frac{3}{2} \ln T + \text{constant} \right) \ . \tag{3.14}$$

This expression may be compared with the entropy of a mole of a monatomic ideal gas according to phenomenological thermodynamics:

$$S_{\text{ideal}}(V, T) = R \left(\ln V + \frac{3}{2} \ln T \right) + \text{constant}' \ , \tag{3.15}$$

with another constant that may depend on the particle number N according to its derivation. The second constant may then be chosen such that

$$S_{\text{ideal}} = -kH[\rho_M] =: S_\mu[\rho_M] \ , \tag{3.16}$$

where $k = R/N$.

The entropy of an ideal gas can thus be identified with the measure of the width of the molecular distribution in μ-space. The *Stoßzahlansatz* successfully describes the evolution of this distribution towards a Maxwell distribution with its parameter T that determines the conserved total energy. This *Lagrange parameter* – see (3.19) – is thereby recognized as the *temperature*.

This important success seems to be the origin of the 'myth' of the statistical foundation of the thermodynamical arrow of time. However, statistical arguments applied to a gas can neither explain why the *Stoßzahlansatz* is a good approximation in one and only one direction of time, nor tell us whether S_μ is *always* an appropriate definition of entropy. It will indeed turn out to be insufficient when correlations between particles become essential, as is the case, for example, for real gases or solid bodies. Taking them into account requires more general concepts, which were first proposed by Gibbs. His approach will also allow us to formulate the exact ensemble dynamics in Γ-space, although it *cannot* yet explain the origin of the thermodynamical arrow of time (that is, of the low-entropy initial conditions).

3.1.2 Γ-Space Dynamics and Gibbs' Entropy

In the preceding section, Boltzmann's *smooth* phase space density ρ_μ was justified by means of small uncertainties in particle positions and momenta. It describes an infinite number (a continuum) of *possible* single-particle states, for example each particle represented by a small volume element ΔV_μ. An objective ('real') state would instead be described by a point (or a δ-distribution) in Γ-space, or by a sum over N δ-functions in μ-space. This would then lead to an infinite value of Boltzmann's H-functional, or negative infinite entropy.

However, the finite value of $S_\mu[\rho_\mu]$, derived from the *smooth* μ-space distribution, is *not* just a measure of this arbitrary smoothing procedure (for example representing the size of the volume elements ΔV_μ). If N points are replaced by small but overlapping volume elements, this leads to a smooth distribution ρ_μ whose width reflects that of the discrete (real) distribution of particles. Therefore, S_μ characterizes the real physical state. The formal 'renormalization of entropy', which is part of this smoothing procedure, adds an infinite positive contribution to the infinite negative entropy corresponding to a point in such a way that the finite result $S_\mu[\rho_\mu]$ is *physically* meaningful. The 'representative ensemble' obtained in this way defines a finite measure of probability (in the sense of the introduction to this chapter) for the $N!$ points in Γ-space. It depends only slightly on the precise smoothing conditions, provided the discrete μ-space distribution is already smooth in the mean.

The ensemble concept introduced by Josiah Willard Gibbs (1902) differs from Boltzmann's at the very outset. He considered probability densities $\rho_\Gamma(p,q)$ with $\int \rho_\Gamma(p,q)\mathrm{d}p\,\mathrm{d}q = 1$ – from now on writing $p := p_1,\ldots,p_{3N}$, $q := q_1,\ldots,q_{3N}$ and $\mathrm{d}p\,\mathrm{d}q := \mathrm{d}^{3N}p\,\mathrm{d}^{3N}q$ for short, which are meant to describe *incomplete information* ('ignorance') about microscopic degrees of freedom. For example, a probability density may characterize a macroscopic (incomplete) preparation procedure. Boltzmann's H-functional is then replaced by Gibbs' formally analogous *extension in phase* η:

$$\eta[\rho_\Gamma] := \overline{\ln \rho_\Gamma} = \int \rho_\Gamma(p,q) \ln \rho_\Gamma(p,q)\mathrm{d}p\,\mathrm{d}q \,. \tag{3.17}$$

It leads generically to a finite *ensemble entropy* $S_\Gamma := -k\eta[\rho_\Gamma]$. For a probability density that is constant on a phase space volume element of size ΔV_Γ (while vanishing elsewhere), one has $\eta[\rho_\Gamma] = -\ln \Delta V_\Gamma$. The entropy $S_\Gamma = k \ln \Delta V_\Gamma$

is a logarithmic measure of the size of this volume element: it does not characterize a real state, as Boltzmann's entropy was supposed to do.

For a smooth distribution of statistically independent particles, $\rho_\Gamma{}' = \prod_{i=1}^{N} \left[\rho_\mu(\boldsymbol{p}_i, \boldsymbol{q}_i)/N\right]$, one nevertheless obtains

$$\eta[\rho_\Gamma] = \sum_{i=1}^{N} \int \left[\rho_\mu(\boldsymbol{p}_i, \boldsymbol{q}_i)/N\right] \ln \left[\rho_\mu(\boldsymbol{p}_i, \boldsymbol{q}_i)/N\right] \mathrm{d}^3 p_i \mathrm{d}^3 q_i$$

$$= \int \rho_\mu(\boldsymbol{p}, \boldsymbol{q}) \left[\ln \rho_\mu(\boldsymbol{p}, \boldsymbol{q}) - \ln N\right] \mathrm{d}^3 p \, \mathrm{d}^3 q = H[\rho_\mu] - N \ln N . \quad (3.18)$$

In this important special case one thus recovers Boltzmann's statistical entropy S_μ (with all its advantages) – except for the term $kN \ln N \approx k \ln N!$ that has to be interpreted as the *mixing entropy* of the gas with itself. It is absent in Boltzmann's approach, since his μ-space distribution does not distinguish between particle permutations even though they define different states. While merely an additive constant in systems with fixed particle number, this self-mixing entropy leads to observable consequences *at variance with experimental results* in situations where the particle number may vary dynamically. Large particle numbers would then acquire far too large statistical weights. In particular, the specific volume V/N in (3.14) would then be replaced by the total volume V. This does even appear consistent (though empirically wrong), since particles forming an ideal gas are independent of one another, so each one is constrained only to the *total* volume V.

Since empirically not required, this self-mixing entropy was generally overlooked in Boltzmann's approach, although it had already been known as a problem to Maxwell. It can be resolved only by applying Gibbs' ensemble concept to quantum states defined in the occupation number representation for field modes (field quantization).[2] Only after borrowing this result from quantum field theory may one identify Boltzmann's entropy with an ensemble entropy (representing incomplete knowledge) for non-interacting 'particles'.

---

[2] The popular argument that this self-mixing entropy has to be dropped simply because of the *indistinguishability* of particles is wrong, since conceptually different (even though operationally indistinguishable) states would have to be counted separately for statistical purposes. Classical states differing by a permutation of particles would dynamically retain their individuality. The use of μ-space distributions, such as in Boltzmann's statistical mechanics, is also inconsistent from a classical point of view, unless these probability densities were multiplied by the weight factors $N!$ again. The concepts of indistinguishability and identity are different in principle (see also Saunders 2005 and references therein for a discussion). The identity of states with interchanged 'particles' can be understood in terms of quantum *fields* – see also (4.21), since the permutation of two identical wave packets at different places would represent an identity operation (Zeh 2003). Even the difference $N \ln N - \ln N! \approx N - \ln N$, usually neglected in these arguments, can be understood: it counts states with different particle numbers which must contribute to open systems that permit particle exchange, described by a *grand canonical* distribution with given chemical potential (see p. 71).

Furthermore, S_Γ is maximized under the constraint of fixed *mean* energy, $\bar{E} = \int H(p,q)\rho_\Gamma(p,q)\mathrm{d}p\,\mathrm{d}q$, by the *canonical* (or Gibbs') distribution $\rho_{\mathrm{can}} := Z^{-1}\exp\left[-H(p,q)/kT\right]$. The latter can be derived from a variational procedure with the additional constraint of fixed normalization of probability, $\int \rho_\Gamma(p,q)\mathrm{d}p\,\mathrm{d}q = 1$, that is, from

$$\delta\left\{\eta[\rho_\Gamma] + \alpha\int\rho_\Gamma(p,q)\mathrm{d}p\,\mathrm{d}q + \beta\int H(p,q)\rho_\Gamma(p,q)\mathrm{d}p\,\mathrm{d}q\right\} \tag{3.19}$$

$$= \int\left[\ln\rho_\Gamma(p,q) + (\alpha+1) + \beta H(p,q)\right]\delta\rho_\Gamma(p,q)\mathrm{d}p\,\mathrm{d}q = 0\,,$$

with Lagrange parameters α and β for fixed normalization and energy. The solution is

$$\rho_{\mathrm{can}} = \exp\left\{-\left[\beta H(p,q) - \alpha - 1\right]\right\} =: Z^{-1}\exp\left[-\beta H(p,q)\right]\,, \tag{3.20}$$

and one recognizes $\beta = 1/kT$ and the *partition function* (sum over states) $Z := \int\mathrm{e}^{-\beta H(p,q)}\mathrm{d}p\,\mathrm{d}q = \mathrm{e}^{-\alpha-1}$. By using the *Ansatz* $\rho = \mathrm{e}^{\chi+\Delta\chi}$ with $\mathrm{e}^\chi := \rho_{\mathrm{can}}$, an arbitrary (not necessarily small) variation $\Delta\chi(p,q)$, the above constraints, and the general inequality $\Delta\chi\mathrm{e}^{\Delta\chi} \geq \Delta\chi$, one may even show that the canonical distribution represents an *absolute* maximum of this entropy. In statistical thermodynamics (and in contrast to phenomenological thermodynamics), entropy is thus a more fundamental concept than temperature, which applies only to special (canonical or equivalent) probability distributions, while a formal entropy is defined for *all* ensembles.

One can similarly show that S_Γ is maximized by the microcanonical ensemble $\rho_{\mathrm{micro}} \equiv \delta(E - H(p,q))$ if constrained by the condition of *fixed energy*, $H(p,q) = E$. Although essentially equivalent for most applications, the canonical and microcanonical distributions characterize two different situations: systems with and without energy exchange with a heat bath.

For non-interacting particles, $H = \sum_i\left[\boldsymbol{p}_i^2/2m + V(\boldsymbol{q}_i)\right]$, one obtains from (3.20) a factorizing canonical distribution $\rho_\Gamma(p,q) = \prod_i\left[\rho_\mu(\boldsymbol{p}_i,\boldsymbol{q}_i)/N\right]$, as already considered in (3.18), with a μ-space distribution given by $\rho_\mu(\boldsymbol{p},\boldsymbol{q}) \propto N\exp\left\{-\left[\boldsymbol{p}^2/2m + V(\boldsymbol{q})\right]/kT\right\}$. This is a Maxwell distribution multiplied by the barometric formula. However, the essential advantage of the canonical Γ-space distribution (3.20) over Boltzmann's is its ability to describe equilibrium correlations between particles. This has been demonstrated in particular by the cluster expansion of Ursell and Mayer (see Mayer and Mayer 1940), in more recent terminology called an expansion by N-point functions, and technically a predecessor of Feynman graphs. However, the distribution (3.20) must not include macroscopic degrees of freedom (such as the position and shape of a solid body). In the case of a rotationally symmetric Hamiltonian, for example, the solid body in thermodynamical equilibrium would otherwise have to be physically characterized by a symmetric distribution of all its orientations in space rather than by a definite orientation. Similarly, its center of mass would always have to be expected close to the minimum of an external potential (see also Fröhlich 1973). These macroscopic variables are *dynamically robust* rather than behaving ergodically. In order to calculate a thermodynamically meaningful representative ensemble according to (3.19), one has to impose additional constraints to fix their values (see Sect. 3.3.1).

Gibbs' extension in phase η thus appears superior to Boltzmann's H-functional (3.9). Unfortunately, the corresponding *ensemble entropy* S_Γ has two (related) defects, which render it entirely unacceptable for representing physical entropy: (1) in stark contrast to the Second Law it remains constant under exact (Hamiltonian) dynamics, and (2) it is obviously not an additive (or extensive) quantity, that would define an entropy *density*.

In order to confirm the first statement, one may formulate the exact ensemble dynamics in Γ-space in analogy to (3.4) by using the $6N$-dimensional continuity equation

$$\frac{\partial \rho_\Gamma}{\partial t} + \mathrm{div}_\Gamma(\rho_\Gamma \boldsymbol{v}_\Gamma) = 0 \; . \tag{3.21}$$

It describes the conservation of probabilities for volume elements moving through Γ-space by forming a bunch of trajectories. The $6N$-dimensional velocity $\boldsymbol{v}_\Gamma$ may be replaced by means of the Hamiltonian equations,

$$\boldsymbol{v}_\Gamma \equiv (\dot{p}_1, \ldots, \dot{p}_{3N}, \dot{q}_1, \ldots, \dot{q}_{3N}) = \left(-\frac{\partial H}{\partial q_1}, \ldots, -\frac{\partial H}{\partial q_{3N}}, \frac{\partial H}{\partial p_1}, \ldots, \frac{\partial H}{\partial p_{3N}} \right) . \tag{3.22}$$

So when rewriting (3.21) by means of the identity

$$\mathrm{div}_\Gamma(\rho_\Gamma \boldsymbol{v}_\Gamma) = \rho_\Gamma \mathrm{div}_\Gamma \boldsymbol{v}_\Gamma + \boldsymbol{v}_\Gamma \cdot \boldsymbol{\nabla}_\Gamma \rho_\Gamma \; ,$$

one may use the *Liouville theorem,*

$$\mathrm{div}_\Gamma \boldsymbol{v}_\Gamma = -\frac{\partial^2 H}{\partial p_1 \partial q_1} - \cdots - \frac{\partial^2 H}{\partial p_{3N} \partial q_{3N}} + \frac{\partial^2 H}{\partial q_1 \partial p_1} + \cdots + \frac{\partial^2 H}{\partial q_{3N} \partial p_{3N}} \equiv 0 \; , \tag{3.23}$$

which describes an incompressible 'fluid' in Γ-space. One thus obtains the *Liouville equation,*

$$\frac{\partial \rho_\Gamma}{\partial t} = -\boldsymbol{v}_\Gamma \cdot \boldsymbol{\nabla}_\Gamma \rho_\Gamma = \sum_{n=1}^{3N} \left(\frac{\partial H}{\partial q_n} \frac{\partial \rho_\Gamma}{\partial p_n} - \frac{\partial H}{\partial p_n} \frac{\partial \rho_\Gamma}{\partial q_n} \right) = \{H, \rho_\Gamma\} \; , \tag{3.24}$$

where $\{a, b\}$ defines the Poisson bracket for two functions a and b. This equation represents the exact Hamiltonian dynamics for ensembles $\rho_\Gamma(p, q, t)$ under the assumption of individually conserved probabilities.

From this analogy with an incompressible fluid in space one may expect the ensemble entropy S_Γ (the measure of 'extension in phase') to remain constant in time. This can indeed be confirmed by differentiating (3.17), inserting (3.24), and repeatedly integrating by parts:

$$\frac{\mathrm{d}S_\Gamma}{\mathrm{d}t} = \int (\ln \rho_\Gamma + 1) \dot{\rho}_\Gamma \mathrm{d}p \, \mathrm{d}q$$

$$= \int (\ln \rho_\Gamma + 1) \sum_{n=1}^{3N} \left(\frac{\partial H}{\partial q_n} \frac{\partial \rho_\Gamma}{\partial p_n} - \frac{\partial H}{\partial p_n} \frac{\partial \rho_\Gamma}{\partial q_n} \right) \mathrm{d}p \, \mathrm{d}q$$

$$= -\int \sum_{n=1}^{3N} \left(\frac{\partial H}{\partial q_n} \frac{\partial \ln \rho_\Gamma}{\partial p_n} - \frac{\partial H}{\partial p_n} \frac{\partial \ln \rho_\Gamma}{\partial q_n} \right) \rho_\Gamma \mathrm{d}p \, \mathrm{d}q$$

$$= -\int \sum_{n=1}^{3N} \left(\frac{\partial H}{\partial q_n} \frac{\partial \rho_\Gamma}{\partial p_n} - \frac{\partial H}{\partial p_n} \frac{\partial \rho_\Gamma}{\partial q_n} \right) \mathrm{d}p \, \mathrm{d}q = 0 \; . \tag{3.25}$$

A more instructive proof may be obtained by multiplying the Liouville equation (3.24) by the imaginary unit i in order to cast the dynamics into the *form* of a Schrödinger equation,

$$i\frac{\partial \rho_\Gamma}{\partial t} = i\{H, \rho_\Gamma\} =: \hat{L}\rho_\Gamma \; . \tag{3.26}$$

The operator $\hat{L}$ (acting on probability densities) is called the *Liouville operator*. In accordance with this analogy one may use the formal solution $\rho_\Gamma(t) = \exp(-i\hat{L}t)\rho_\Gamma(0)$, valid if $\partial \hat{L}/\partial t = 0$ (see Prigogine 1962). The Liouville operator is Hermitean with respect to the inner product $\langle \rho_\Gamma, \rho_\Gamma' \rangle := \int \rho_\Gamma^* \rho_\Gamma' \mathrm{d}p \, \mathrm{d}q$ (that is, $\langle \rho_\Gamma, \hat{L}\rho_\Gamma' \rangle = \langle \hat{L}\rho_\Gamma, \rho_\Gamma' \rangle$), as can again be shown by partial integration. This means that the Liouville equation conserves these inner products. In particular, for $\rho_\Gamma' = \ln \rho_\Gamma$, one has

$$\frac{\mathrm{d}}{\mathrm{d}t}\langle \rho_\Gamma, \ln \rho_\Gamma \rangle = \frac{\mathrm{d}}{\mathrm{d}t}\overline{\ln \rho_\Gamma} = 0 \; , \tag{3.27}$$

since the Liouville operator, when applied to a function $f(\rho_\Gamma)$, satisfies the same Leibniz chain rule $\hat{L}f(\rho_\Gamma) = (\mathrm{d}f/\mathrm{d}\rho_\Gamma)\hat{L}\rho_\Gamma$ as the time derivative.

The *norm* corresponding to this inner product, $\|\rho_\Gamma\|^2 = \langle \rho_\Gamma, \rho_\Gamma \rangle = \int \rho_\Gamma^2 \mathrm{d}p \, \mathrm{d}q = \overline{\rho_\Gamma}$, is then also dynamically invariant. It represents a linear measure of extension in phase (a *linear ensemble entropy*[3]), and thus has to be distinguished from the probability norm $\int \rho_\Gamma \mathrm{d}p \, \mathrm{d}q = \bar{1} = 1$. The conservation of these measures under a Liouville equation confirms in turn that the Γ-space volume is an appropriate measure for non-countable sets of states (Ehrenfest and Ehrenfest 1911): the thus defined 'number' of states does not change under an appropriately defined determinism. A more fundamental justification of this measure can be derived from the conservation of probabilities of discrete *quantum* states (see Sect. 4.1).

---

[3] See Wehrl (1978) for further measures, which are, however, not always monotonically related to one another. The conventional logarithmic measure is usually preferred because of the resulting additivity of the entropies of statistically independent subsystems.

The conservation of ensemble entropy, implied by using the exact dynamics, is unacceptable in a statistical foundation of *physical* entropy. Therefore, Gibbs introduced a more subtle concept of entropy, that was motivated by his famous *ink drop analogy*: A bit of ink dropped into a glass of water is assumed to behave as an incompressible fluid when the water is stirred. Although its volume must remain constant, the whole glass of water will soon appear homogeneous in light blue. Only a microscopic examination would reveal that the ink had simply rearranged itself into many thin tubes, which are everywhere dense in spite of occupying only a volume of the initial size of the droplet.

Therefore, Gibbs defined his new entropy S_{Gibbs} by means of a *coarse-grained* distribution ρ^{cg}, obtained by averaging over small (but fixed) $6N$-dimensional volume elements ΔV_m ($m = 1, 2, \ldots$) which cover the whole Γ-space:

$$\rho^{\text{cg}}(p, q) = \frac{1}{\Delta V_m} \int_{\Delta V_m} \rho(p', q') \mathrm{d}p' \mathrm{d}q' =: \frac{\Delta p_m}{\Delta V_m} , \quad \text{for } p, q \in \Delta V_m . \quad (3.28)$$

The resulting ensemble entropy is then given by

$$S_{\text{Gibbs}} := -k\eta[\rho^{\text{cg}}] = -k \sum_m \Delta p_m \ln \frac{\Delta p_m}{\Delta V_m} . \quad (3.29)$$

As already mentioned in connection with the smoothing of Boltzmann's μ-space distributions, the justification of this procedure by means of the quantum mechanical uncertainty relations, that is, by coarse-graining over phase space cells of size h^{3N}, may be tempting, but would clearly be inconsistent with classical mechanics. The consistent quantum mechanical treatment (Chap. 4) leads again to the conservation of ensemble entropy (now for ensembles of wave functions rather than Γ-space points). 'Quantum cells' of size h^{3N} can be justified only as convenient *units* of phase space volume in order to obtain the same normalization of entropy as in the classical limit of *quantum* statistical mechanics, where ensemble entropy vanishes for pure states, which correspond to phase space 'cells' – see (4.21). However, these quantum cells do *not* define uncertain initial conditions which might explain quantum indeterminism (as often claimed); ensemble entropy is conserved under Hamiltonian *and* Schrödinger dynamics.

The increase in Gibbs' entropy can be understood according to the classical ink drop analogy. While the volume of the compact ink droplet is only slightly increased by moderate coarse-graining, that of a dense web of thin tubes (obtained by stirring) is considerably enlarged. Even though the coarse-graining itself is artificial, its efficiency depends on the shape of the volume to which it is applied. This is similar to using Boltzmann's smooth μ-space densities, which characterize properties of the *discrete* particle distributions. Since there evidently exist far more droplet shapes with a large surface than compact ones, the former have to be regarded as more probable. For *statistical* reasons one should hardly ever find a compact droplet (which is confirmed for three-dimensional 'droplets' in the absence of any surface tension).

However, there remains an essential difference between a droplet of ink in water and a dynamical volume element in phase space. While the extension and shape of a droplet are real physical properties, the real state of a classical mechanical system is represented by a *point* in phase space. Coarse-graining of the ink drop may be likened to Boltzmann's smoothing procedure in-so-far as it preserves properties of the discrete particle distribution, while Gibbs' entropy for a real state p, q, $S_{\text{Gibbs}} = f(p, q) := k \ln \Delta V_{m_0}$ (resulting if $p, q \in \Delta V_{m_0}$), would be entirely artificial. This difference would be reduced if individual classical state were identified with $N!$ points in $\Gamma-$space.

Gibbs' procedure is therefore usually applied to presumed phase space densities, which can only represent incomplete *information*. His entropy then measures the enlargeability by coarse-graining of a certain state of knowledge – not by coarse-graining of a real physical state. Its increase, $dS_{\text{Gibbs}}/dt \geq 0$, under a deterministic (information-conserving) dynamical law describes the transformation of macroscopic information, assumed to be present *initially*, into fine-grained information, that is then regarded as 'irrelevant' and dynamically neglected (Sect. 3.2). However, this procedure may be in conflict with the idea of entropy as an objective physical quantity that is independent of any information held by an observer. This fundamental problem will be addressed again in Sect. 3.3 and later chapters.

Similar to the problem that arose for μ-space densities in (3.2), the coarse-graining cannot be uniquely inverted, since it destroys information. The intended chain of calculation,

$$\rho^{\text{cg}} \longrightarrow \rho \xrightarrow{\hat{L}} \frac{\partial \rho}{\partial t} \longrightarrow \frac{\partial \rho^{\text{cg}}}{\partial t}, \tag{3.30}$$

is again broken at its first link. A new autonomous dynamics has therefore been proposed for ρ^{cg}, in analogy to the *Stoßzahlansatz*, by complementing the Hamiltonian dynamics with a *dynamical* coarse-graining, applied in small but finite time steps Δt:

$$\left\{ \frac{\partial \rho^{\text{cg}}}{\partial t} \right\}_{\text{master}} := \frac{\left[e^{-i\hat{L}\Delta t} \rho^{\text{cg}} \right]^{\text{cg}} - \rho^{\text{cg}}}{\Delta t}. \tag{3.31}$$

In this form it may also be regarded as a variant of a 'unifying principle' that was proposed as a stochastic process by R.M. Lewis (1967). Instead of dynamically applying Gibbs' coarse-graining in (3.31), Lewis suggested *maximizing the entropy* in each dynamical step under the constraint of certain fixed 'macroscopic' quantities (see also Jaynes' theory in Sect. 3.3.1).

Equation (3.31) defines reasonable dynamics if the corresponding probability increments $\Delta\Delta p_m$ – see (3.28) – are proportional to Δt for small but finite time intervals Δt, thus describing transition *rates* between the cells ΔV_m. This important condition will be discussed in a more general form in Sect. 3.2, and later for deriving the Pauli equation (4.18). Master equations such as (3.31) ensure a monotonic entropy increase. Their approximate validity requires that

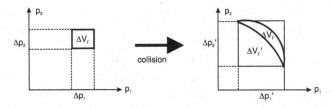

Fig. 3.1. Transformation of information about particle momenta into information about correlations between them as the basis of the H-theorem (symbolic, for non-central collisions)

the arising microscopic (fine-grained) information remains *dynamically* irrelevant for the evolution of the coarse-grained distribution. Except in the case of equilibrium, this cannot simultaneously be true in different directions of time.

The meaning of Boltzmann's *Stoßzahlansatz* (3.6) can be similarly understood, as it neglects all particle correlations, which are thus regarded as fine-grained information, *after* they have formed in collisions. It is again based on the assumption that the interval Δt is finite and large compared to collision times. The effect of an individual collision on the phase space distribution may be illustrated in two-dimensional momentum space (Fig. 3.1): a collision between two particles with small momentum uncertainties Δp_1 and Δp_2 leads deterministically to a *correlating* (deformed) volume element of the same size ΔV_Γ as the initial one. (In a realistic description, momenta would also be correlated with particle positions.) Subsequent neglect of the arising correlations will then enlarge this volume element ($\Delta V_\Gamma' > \Delta V_\Gamma$). However, neglecting such statistical correlations evidently has no effect on a phase space *point*.

The question as to the precise mathematical conditions under which certain systems are indeed 'mixing' in the sense of the plausible ink drop analogy (in a stronger version referred to as *K-systems* after Kolmogorov) is rigorously investigated, though under idealized conditions (such as ideal isolation), in *ergodic theory* (see Arnol'd and Avez 1968, or Mackey 1989). Most non-ergodic systems are pathological in forming sets of measure zero, or in being unstable against unavoidable perturbations. In general, the quantitative question for the *time-scale* of mixing between different regions in phase space is physically far more relevant than exact formal theorems which apply only at infinite times. Regions which don't mix with others over long times may define robust (usually macroscopic) properties, that do *not* have to represent constants of the motion. On the other hand, certain non-ergodic aspects have been claimed to apply under quite general circumstances (Yoccoz 1992), but no physically relevant interpretation of these formal dynamical properties has ever been given.

The strongest mixing is required for the finest conceivable coarse-graining. This is given by its nontrivial limit $\Delta V_\Gamma \to 0$ for the size of grains, which defines a *weak convergence* for measures on phase space. It would again lead to infinite Poincaré recurrence times for isolated systems. However, this is neither required in a universe of finite age, nor would it be realistic, since quantum theory limits the entropy capacity available in the form of unlimited fine-graining of classical phase space. For this quantum mechanical reason there can be no 'overdetermination' of the *microscopic* past in spite of the validity of microscopic causality (see footnote 1 of Chap. 2 and the end of Sect. 5.3). However, it is important to note that all concepts of mixing are T-symmetric. In order to explain the time asymmetry of the Second Law ('irreversibility'), they would have to be *applied dynamically* in a specific direction of time.

Dynamical coarse-graining as in (3.31) may also be based on an incompletely known Hamiltonian. An *ensemble of Hamiltonians* defines a stochastic dynamical model when used for calculating 'forward' in time. Even very small uncertainties in the Hamiltonian may be sufficient to completely destroy fine-grained information within a short time interval. Borel (1924) estimated the effect of a gravitational force that would arise here on earth by the displacement of a mass of the order of a few grams by a few centimeters at the distance of Sirius. He thereby pointed out that this would lead to a completely different microscopic state for the molecules forming a gas in a vessel under normal conditions within seconds. Although distortions of the individual molecular trajectories are extremely small, they would be amplified in each subsequent collision by a factor of the order of l/R, the ratio of the mean free path over the molecular radius. This extreme sensitivity to the environment describes in effect a *local microscopic indeterminism*.[4] In many situations, the microscopic distortions may even co-determine macroscopic effects (thus inducing an effective *macroscopic indeterminism*), as discussed, in particular, in the *theory of chaos* ('butterfly effect').

The essence of Borel's argument is that macroscopic systems, aside from the whole Universe, may never be regarded as dynamically isolated – even when thermodynamically closed in the sense of $dS_{ext} = 0$. The *dynamical* coarse-graining that is part of the master equation (3.31) may indeed be ascribed to perturbations by the environment – provided the latter obey causality, that is, can be treated stochastically in the forward direction of time. This important dynamical assumption is yet another form of the *intuitive causality* discussed at the beginning of Chap. 2 as a major manifestation of the arrow of time. The *representative ensembles* used in statistical thermodynamics may therefore be understood within classical physics as those which arise (and are maintained) by this stochastic nature of unavoidable perturbations, while 'robust' properties can be regarded as macroscopic.

While the *intrinsic dynamics* of a macroscopic physical system transforms coarse-grained into fine-grained information, interactions with the environment thus transform the resulting fine-grained information very efficiently into practically useless correlations with distant systems. The sensitivity of

---

[4] While the effect of Borel's gravitational distortion is drastically reduced for quantized interactions, other environmental effects (such as decoherence) then become important in producing an effective local indeterminism (see Sects. 4.3.4 and 5.3).

the microscopic states of macroscopic systems to such interactions with their environments strongly indicates that simultaneously existing *opposite arrows of time* in different regions of the Universe would be inconsistent with one another. This universality of the arrow of time seems to be its most important property. Time asymmetry has therefore been regarded as a global *symmetry breaking*. However, such a conclusion would *not* exclude the far more probable situation of thermal equilibrium.

Lawrence Schulman (1999) has challenged the usual assumption of a universal arrow of time by suggesting explicit counterexamples. Most of them are indeed quite illustrative in emphasizing the role of initial of final conditions, but they appear unrealistic in our Universe (see Zeh 2005b). The situation is similar to the symmetric boundary conditions suggested by Wheeler and Feynman in electrodynamics, and discussed in Sect. 2.4. Local final conditions at the present stage of the Universe or in the near future can hardly be retro-caused by a low entropy condition at the big crunch (see also Casati, Chirikov and Zhirov 2000), but may be essential during a conceivable recontraction era of the Universe (see Sect. 5.3).

In order to reverse the thermodynamical arrow of time in a bounded system, it would not therefore suffice to "go ahead and reverse all momenta" in the system itself, as ironically suggested by Boltzmann as an answer to Loschmidt. In an interacting Laplacean universe, the Poincaré cycles of its subsystems could in general only be those of the whole Universe, since their exact Hamiltonians must always depend on their time-dependent environment.

Time reversal including thermodynamical aspects has been achieved even in practice for very weakly interacting spin waves (Rhim, Pines and Waugh 1971). The latter can be regarded as isolated systems to a very good approximation (similar to electromagnetic waves in the absence of absorbers), while allowing a sudden sign reversal of their spinor Hamiltonian in order to simulate time reversal $(dt \rightarrow -dt)$. These spin wave experiments demonstrate that a closed system in thermodynamical equilibrium may preserve an arrow of time in the form of *hidden correlations*. When a closed system has reached macroscopic equilibrium, it *appears* T-symmetric, although its fine-grained information determines the distance and direction in time to its low-entropy state in the past (see also the Appendix for a numerical example). In contrast to such rare almost-closed systems, generic ones are strongly affected by Borel's argument, and *cannot* be reversed by local manipulations.

3.2 Zwanzig's General Formalism of Master Equations

Boltzmann's *Stoßzahlansatz* (3.6) for μ-space distributions and the master equation (3.31) for coarse-grained Γ-space distributions can thus be understood in a similar way. They describe the transformation of *special* macroscopic states into more probable ones, whereby the higher information content of the former is transformed into macroscopically irrelevant information. There are many other master equations based on the same strategy, and designed to suit various purposes. Zwanzig (1960) succeeded in formalizing them in a general and instructive manner that also reveals their analogy with retarded electrodynamics as another manifestation of the arrow of time – see (3.40)–(3.49) below.

The basic concept of Zwanzig's formalism is defined by idempotent mappings $\hat{P}$, acting on probability distributions $\rho(p, q)$:

$$\rho \rightarrow \rho_{\text{rel}} := \hat{P}\rho \,, \quad \text{with} \quad \hat{P}^2 = \hat{P} \quad \text{and} \quad \rho_{\text{irrel}} := (1 - \hat{P})\rho \,. \quad (3.32)$$

Their meaning will be illustrated by means of several examples below, before explaining the dynamical formalism. If these mappings *reduce* the information content of ρ to what is then called its 'relevant' part ρ_{rel}, they may be regarded as a *generalized coarse-graining*. In order to interpret ρ_{rel} as a probability density again, one has to require its non-negativity and, for convenience,

$$\int \rho_{\text{rel}} \mathrm{d}p \, \mathrm{d}q = \int \rho \mathrm{d}p \, \mathrm{d}q = 1 \,, \quad (3.33)$$

that is,

$$\int \rho_{\text{irrel}} \mathrm{d}p \, \mathrm{d}q = \int (1 - \hat{P})\rho \mathrm{d}p \, \mathrm{d}q = 0 \,. \quad (3.34)$$

Reduction of information means

$$S_\Gamma[\hat{P}\rho] \geq S_\Gamma[\rho] \quad (3.35)$$

(or similarly for any other measure of ensemble entropy).

Using this concept, Lewis' master equation (3.31), for example, may be written in the generalized form

$$\left\{ \frac{\partial \rho_{\text{rel}}}{\partial t} \right\}_{\text{master}} := \frac{\hat{P}\mathrm{e}^{-\mathrm{i}\hat{L}\Delta t}\rho_{\text{rel}} - \rho_{\text{rel}}}{\Delta t} \,. \quad (3.36)$$

It would then describe a monotonic increase in the corresponding entropy $S[\rho_{\text{rel}}]$. In contrast to Zwanzig's approach, to be described below, phenomenological master equations such as Lewis's unifying principle have often been meant to describe a *fundamental* indeterminism that would replace reversible Laplacean determinism.

In most applications, Zwanzig's idempotent operations $\hat{P}$ are linear and Hermitean with respect to the inner product for probability distributions defined above (3.27). In this case they are projection operators, which preserve only some 'relevant component' of the original information. If such a projection obeys (3.33) for every ρ, it must leave the equipartition invariant, $\hat{P}1 = 1$, as can be shown by writing down the above-mentioned inner product of this equation with an *arbitrary* distribution ρ and using the hermiticity of $\hat{P}$.

Zwanzig's dynamical formalism may also be useful for non-Hermitean or even non-linear idempotent mappings $\hat{P}$ (see Lewis 1967, Willis and Picard 1974). These mappings are then not projections any more: they may even *create new information*. A trivial example for the creation of information is the nonlinear mapping of all probability distributions onto a fixed one, $\hat{P}\rho := \rho_0$ for all ρ, regardless of whether or not they contain a component proportional to ρ_0. The physical meaning of such generalizations of Zwanzig's formalism will be discussed in Sects. 3.4 and 4.4. In the following we shall consider information-reducing mappings.

Zwanzig's 'projection' concept is deliberately kept general in order to permit a wealth of applications. Examples introduced so far are coarse-graining, $\hat{P}_{cg}\rho := \rho^{cg}$, as defined in (3.28), and the neglect of correlations between particles by means of μ-space densities:

$$\hat{P}_\mu\rho(p,q) := \prod_{i=1}^{N} \frac{\rho_\mu(\boldsymbol{p}_i, \boldsymbol{q}_i)}{N} ,$$

with

$$\rho_\mu(\boldsymbol{p}, \boldsymbol{q}) := \sum_{i=1}^{N} \int \rho(p,q)\delta^3(\boldsymbol{p} - \boldsymbol{p}_i)\delta^3(\boldsymbol{q} - \boldsymbol{q}_i)\mathrm{d}p\,\mathrm{d}q . \tag{3.37}$$

(As before, boldface letters represent three-dimensional vectors, while p, q is a point in Γ-space.) The latter example defines a *non-linear* though information-reducing 'Zwanzig projection'. Most arguments applying to linear operators $\hat{P}$ remain valid in this case when applied to the linearly resulting μ-space distributions $\rho_\mu(\boldsymbol{p}, \boldsymbol{q})$ (which do not live in Γ-space) rather than to their products $\hat{P}_\mu\rho(p,q)$ (which do). In quantum theory, this approach is related to the Hartree or mean field approximation. Boltzmann's 'relevance concept', which, when written as a Zwanzig projection, would map real states onto products of *smooth* μ-space distributions, can then be written as $\hat{P}_{Boltzmann} = \hat{P}_\mu\hat{P}_{cg}$. An obvious generalization of $\hat{P}_{Boltzmann}$ can be defined by a projection onto two-particle correlation functions. In this way, a complete hierarchy of relevance concepts in terms of *n-point functions* (equivalent to a cluster expansion) can be defined.

A particularly important concept of relevance, that is often not even noticed, is *locality* (see, e.g., Penrose and Percival 1962). It is required in order to define entropy as an extensive quantity – in accordance with the phenomenological equation (3.1) and with the concept of an entropy *density* $s(\boldsymbol{r})$, such that $S = \int s(\boldsymbol{r})\mathrm{d}^3r$. The corresponding Zwanzig projection of locality may be symbolically written as

$$\hat{P}_{local}\rho := \prod_k \rho_{\Delta V_k} . \tag{3.38}$$

The RHS here is meant to describe the neglect of all statistical correlations beyond a distance defined by the size of volume elements ΔV_k. The probability distributions $\rho_{\Delta V_k}$ would here be defined by integrating over all external degrees of freedom. The volume elements have to be chosen large enough to contain a sufficient number of particles in order to preserve dynamically relevant short range correlations (as required for real gases, for example). In order to allow volume elements ΔV_k with physically open boundaries, their probability distributions $\rho_{\Delta V_k}$ in (3.38) have to admit variable particle number (density fluctuations) – as in a grand canonical ensemble.

Locality is presumed, in particular, when writing (3.1) in its differential (local) form as a 'continuity inequality' for the entropy density $s(r, t)$,

$$\frac{\partial s}{\partial t} + \mathrm{div} \boldsymbol{j}_s \geq 0 \ , \tag{3.39}$$

with an entropy current density $\boldsymbol{j}_s(r, t)$. This form allows the definition of phenomenological entropy-producing (hence positive) terms on the RHS in order to replace the inequality by an *equation* (see Landau and Lifschitz 1959 or Glansdorff and Prigogine 1971). An example is the source term $\kappa (\nabla T)^2 / T^2$ in the case of heat conduction, where κ is the heat conductivity.

The general applicability of (3.39) demonstrates that the concept of *physical entropy* is always based on the neglect of nonlocal correlations. Therefore, the production of entropy can be usually understood as the transformation of local information into nonlocal correlations (as depicted in Fig. 3.1). This description is in accordance with the conservation of ensemble entropy (determinism) *and* with intuitive causality. The Second Law thus depends crucially on the dynamical irrelevance of microscopic correlations *for the future* (as assumed in the *Stoßzahlansatz*, for example). Since this 'microscopic causality' cannot be observed as easily and directly as the causal correlations which define retardation of macroscopic radiation, its validity under all circumstances has been questioned (Price 1996). However, it is not only indirectly confirmed by the success of the *Stoßzahlansatz*, but also (in its quantum mechanical form – see Sect. 4.2) by the validity of a Sommerfeld radiation condition (see Sect. 2.1) for microscopic scattering experiments, or by the validity of exponential decay (Sect. 4.5).

The Zwanzig projection of locality is again ineffective on real states, which are always local in the sense of defining the states of all their subsystems. Therefore, applying $\hat{P}_{local}$ to an individual state (a δ-function or sum of them) would not lead to a non-singular entropy S_Γ. This will drastically change in quantum mechanics, because it is kinematically non-local (Chap. 4).

As already mentioned on p. 55, coarse-graining as a relevance concept may also enter in a hidden form, corresponding to its nontrivial limit $\Delta V_\Gamma \to 0$, by considering only *non-singular measures* on phase space (thus excluding δ-functions). This strong idealization may be mathematically signalled by the 'unitary inequivalence' of the original Liouville equation and the master equations resulting in this limit (see Misra 1978 or Mackey 1989).

Further examples of Zwanzig projections will be defined throughout the book, in particular in Chap. 4 for quantum mechanical applications, where the relevance of locality leads to the important concept of decoherence. Different schools and methods of irreversible thermodynamics may even be distinguished according to the concepts of relevance which they are using, and which they typically regard as 'natural' or 'fundamental' (see Grad 1961).

However, the mere *conceptual* foundation of a relevance concept ('paying attention' only to certain aspects) is insufficient for justifying its *dynamical* autonomy in the form of a master equation (3.36) – see the Appendix for an explicit example. Locality *is* usually dynamically relevant in this sense because of the locality of all interactions. This dynamical locality is essential even for the very concept of physical *systems*, including those of local observers as the ultimate referees for what is relevant.

Zwanzig reformulated the exact Hamiltonian dynamics for $\rho_{\rm rel}$ regardless of any specific choice of $\hat{P}$ instead of simply *postulating* a phenomenological master equation (3.36) in analogy to Boltzmann or Lewis. It can then in general not be autonomous[5], that is, of the form $\partial \rho_{\rm rel}/\partial t = f(\rho_{\rm rel})$, but has to be written as

$$\frac{\partial \rho_{\rm rel}}{\partial t} = f(\rho_{\rm rel}, \rho_{\rm irrel}) \qquad (3.40)$$

in order to eliminate $\rho_{\rm irrel}$ by means of certain assumptions. The procedure is analogous to the elimination of the electromagnetic degrees of freedom by means of the condition $A^\mu_{\rm in} = 0$ when deriving a retarded action-at-a-distance theory (Sect. 2.2). In both cases, empirically justified boundary conditions which specify a time direction are assumed to hold for the degrees of freedom that are to be eliminated.

To this end the Liouville equation $i\partial \rho/\partial t = \hat{L}\rho$ is decomposed into its relevant and irrelevant parts by multiplying it by $\hat{P}$ or $1 - \hat{P}$, respectively:

$$i\frac{\partial \rho_{\rm rel}}{\partial t} = \hat{P}\hat{L}\rho_{\rm rel} + \hat{P}\hat{L}\rho_{\rm irrel} \,, \qquad (3.41a)$$

$$i\frac{\partial \rho_{\rm irrel}}{\partial t} = (1 - \hat{P})\hat{L}\rho_{\rm rel} + (1 - \hat{P})\hat{L}\rho_{\rm irrel} \,. \qquad (3.41b)$$

This corresponds to representing the Liouville operator by a matrix of operators

$$\hat{L} = \begin{pmatrix} \hat{P}\hat{L}\hat{P} & \hat{P}\hat{L}(1 - \hat{P}) \\ (1 - \hat{P})\hat{L}\hat{P} & (1 - \hat{P})\hat{L}(1 - \hat{P}) \end{pmatrix} \,. \qquad (3.42)$$

Equation (3.41b) for $\rho_{\rm irrel}$, with $(1 - \hat{P})\hat{L}\rho_{\rm rel}$ regarded as an inhomogeneity, may then be formally solved by the method of the variation of constants (interaction representation). This leads to

$$\rho_{\rm irrel}(t) = {\rm e}^{-{\rm i}(1-\hat{P})\hat{L}(t-t_0)}\rho_{\rm irrel}(t_0) - {\rm i}\int_0^{t-t_0} {\rm e}^{-{\rm i}(1-\hat{P})\hat{L}\tau}(1 - \hat{P})\hat{L}\rho_{\rm rel}(t - \tau){\rm d}\tau \,, \qquad (3.43)$$

---

[5] In mathematical physics, 'autonomous dynamics' is often defined as the absence of any explicit time dependence in the dynamics – regardless of whether it is fundamental or caused by a time-dependent environment.

as may be confirmed by differentiation.

If $t > t_0$, (3.43) is analogous to the *retarded form* (2.9) of the boundary value problem in electrodynamics. In this case, $\tau \geq 0$, and $\rho_{rel}(t - \tau)$ may be interpreted as an advanced source for the 'retarded' $\rho_{irrel}(t)$. Substituting this formal solution (3.43) into (3.41a) leads to three terms on the RHS, viz.,

$$i\frac{\partial \rho_{rel}(t)}{\partial t} = I + II + III \tag{3.44}$$

$$\equiv \hat{P}\hat{L}\rho_{rel}(t) + \hat{P}\hat{L}e^{-i(1-\hat{P})\hat{L}(t-t_0)}\rho_{irrel}(t_0) - i\int_0^{t-t_0} \hat{G}(\tau)\rho_{rel}(t-\tau)d\tau .$$

The integral kernel of the last term,

$$\hat{G}(\tau) := \hat{P}\hat{L}e^{-i(1-\hat{P})\hat{L}\tau}(1 - \hat{P})\hat{L}\hat{P} , \tag{3.45}$$

corresponds to the retarded Green's function of Sect. 2.1.

Equation (3.44) is exact and, therefore, cannot yet describe time asymmetric dynamics. Since it forms the first step in this derivation of master equations, it is known as a *pre-master equation*. The meanings of its three terms are illustrated in Fig. 3.2. The first one describes the internal dynamics of ρ_{rel}. In Boltzmann's μ-space dynamics (3.3), it would correspond to $\{\partial\rho_\mu/\partial t\}_{free+ext}$. It vanishes if $\hat{P}\hat{L}\hat{P} = 0$ (as is often the case).[6]

The second term of (3.44) is usually omitted by presuming the absence of irrelevant initial information: $\rho_{irrel}(t_0) = 0$. If *relevant* information happens to be present initially, it can then be dynamically transformed into irrelevant information. (Because of the asymmetry between $\hat{P}$ and $1 - \hat{P}$, irrelevant information would have to be measured by $-S_\Gamma[\rho] + S_\Gamma[\rho_{rel}]$ rather than by $-S_\Gamma[\rho_{irrel}]$.)

The vital third term is *non-Markovian* (non-local in time), as it depends on the whole time interval between t_0 and t. Its retarded form (valid for $t > t_0$) is compatible with the intuitive concept of causality. This term becomes approximately *Markovian* if $\rho_{rel}(t - \tau)$ varies slowly for a small 're-laxation time' τ_0 during which $\hat{G}(\tau)$ becomes negligible for reasons to be discussed. In (3.44), $\hat{G}(\tau)$ may then be regarded to lowest order as being proportional to a δ-function in τ. This assumption is also contained in Boltzmann's *Stoßzahlansatz*, where it means that correlations arising by scattering

---

[6] Since the (indirectly acting) non-trivial terms contribute only in second and higher orders of time, the time derivative defined by the master equation (3.36) would then vanish in the limit $\Delta t \rightarrow 0$. This corresponds to what in quantum theory is known as the *quantum Zeno paradox* (Misra and Sudarshan 1977), also called *watched pot behavior* or the *watchdog effect*. It describes an *immediate* loss of information from the irrelevant channel (or its dynamically relevant parts – see later in the discussion), such that it has no chance of affecting its relevant counterpart any more. Fast information loss may be caused by a strong coupling to the environment, for example. Since this efficiency depends on the energy level density (Joos 1984), the Zeno effect is relevant mainly in quantum theory.

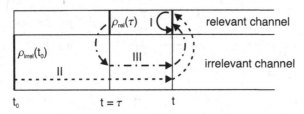

Fig. 3.2. Retarded form of the exact dynamics for the relevant information according to Zwanzig's *pre-master equation* (3.24). In addition to the instantaneous direct interaction I, there is the contribution II arising from the 'incoming' irrelevant information, and the retarded term III in analogy to electromagnetic action at a distance, resulting from 'advanced sources' in the whole time interval between t_0 and t (cf. the left part of Fig. 2.2)

are irrelevant for the forward dynamics of ρ_{rel}. In analogy to retarded electromagnetic forces, this third term of the pre-master equation then assumes the form of an effective direct interaction between the relevant degrees of freedom (though instantaneous in this nonrelativistic treatment). In electrodynamics, the charged sources would represent the 'relevant' variables, while their effective interactions act 'at a distance'. In statistical physics, this 'interaction' describes the dynamics of *ensembles*.

The Markovian approximation may be understood by means of assumptions which simultaneously explain the applicability of the initial condition $\rho_{\mathrm{irrel}} \approx 0$ at *all* times – provided it holds in an appropriate form in the very distant past. This is again analogous to the condition in electrodynamics that A_{in}^μ either vanishes or can be well understood in terms of a limited number of known or at least plausible sources *at all times*.

Consider the action of the operator $(1 - \hat{P})\hat{L}\hat{P}$ appearing on the RHS of the kernel (3.45). Because of the structure of a typical Liouville operator, it transforms information from ρ_{rel} only into *specific* parts of ρ_{irrel}. In the scattering theory of complex objects, similar formal parts are called *doorway states* (Feshbach 1962). For example, if the Hamiltonian contains no more than two-particle interactions, $\hat{L}\hat{P}_\mu$ creates two-particle correlations. Only the subsequent application of the propagator $\exp[-\mathrm{i}(1 - \hat{P})\hat{L}\tau]$ is then able to produce states 'deeper' in the irrelevant channel (many-particle correlations in this case) – see Fig. 3.3. Recurrence from the depth of the irrelevant channel is related to Poincaré recurrence times, and may in general be neglected (as exemplified by the success of Boltzmann's collision equation). If the *relaxation time*, now defined as the time required for the transfer of information from the doorway 'states' into deeper parts of the irrelevant channel, is of the order τ_0, say, one may assume $\hat{G}(\tau) \approx 0$ for $\tau \gg \tau_0$, as required for the Markovian δ-function approximation $\hat{G}(\tau) \approx \hat{G}_0 \delta(\tau)$.

Essential for the validity of this approximation is the large information capacity of the irrelevant channel (similar to that of the electromagnetic field

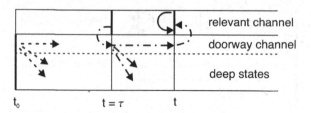

Fig. 3.3. The large information capacity of the irrelevant channel and the specific structure of the interaction together enforce the disappearance of information into the depth of the irrelevant channel if an appropriate initial condition holds

in Chap. 2, but far exceeding it). For example, correlations between particles may describe far more information than the single-particle distribution ρ_μ. A fundamental *cosmological* assumption,

$$\rho_{\mathrm{irrel}}(t_0) = 0 \,, \tag{3.46}$$

at a time t_0 in the *finite* past (similar to the cosmological $A_{\mathrm{in}}^\mu = 0$ at the big bang) is therefore quite powerful – even though it is a *probable* condition. Any irrelevant information formed later from the initial 'information' contained in $\rho_{\mathrm{rel}}(t_0)$ (that is, from any specification of the initial state) may be expected to remain dynamically negligible in (3.44) for a very long time. It would be essential, however, for calculating backwards in time under these conditions.

The assumption $\rho_{\mathrm{irrel}} \approx 0$ has thus to be understood in a *dynamical* sense: any newly formed contribution to ρ_{irrel} must remain irrelevant in the 'forward' direction of time. The dynamics for ρ_{rel} may then appear autonomous (while it cannot be exact). For example, all correlations between subsystems seem to require advanced local *causes*, but no similar (retarded) *effects*. Otherwise they would be interpreted as a *conspiracy*, the deterministic version of *causae finales*.

Under these *assumptions*, one obtains from (3.44), as a first step, the non-Markovian dynamics

$$\frac{\partial \rho_{\mathrm{rel}}(t)}{\partial t} = -\int_0^{t-t_0} \hat{G}(\tau)\rho_{\mathrm{rel}}(t-\tau)\mathrm{d}\tau \,. \tag{3.47}$$

The upper boundary of the integral can here be replaced by a constant T that is large compared to τ_0, but small compared to any (theoretical) recurrence time for $\hat{G}(\tau)$. If $\rho_{\mathrm{rel}}(t)$ is now assumed to remain constant over time intervals of the order of the relaxation time τ_0, corresponding to an already prevailing partial (e.g., local) equilibrium, one obtains the time-asymmetric Markovian limit:

$$\frac{\partial \rho_{\mathrm{rel}}(t)}{\partial t} \approx -\hat{G}_{\mathrm{ret}}\rho_{\mathrm{rel}}(t) \,, \tag{3.48}$$

with

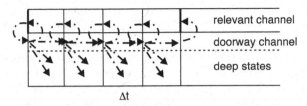

Fig. 3.4. The master equation represents 'alternating dynamics', usually describing a monotonic loss of relevant information

$$\hat{G}_{\text{ret}} := \int_0^T \hat{G}(\tau)\mathrm{d}\tau \ . \tag{3.49}$$

A similar nontrivial limit of vanishing retardation ($\tau_0 \to +0$) led to the LAD equation with its asymmetric radiation reaction in Sect. 2.3. The integral (3.49) could be formally evaluated when inserting (3.45), but it is usually more conveniently computed after this operator has been applied to a specific $\rho(t)$. (See the explicit evaluation for discrete quantum mechanical states in Sect. 4.1.2.)

The autonomous master equation (3.48) again describes *alternating dynamics* of the type (3.36) (see Fig. 3.4). Irrelevant information is disregarded after short time intervals Δt (now representing the relaxation time τ_0). If $\hat{P}$ only *destroys* information, the master equation describes never-decreasing entropy:

$$\frac{\mathrm{d}S_\Gamma[\rho_{\text{rel}}]}{\mathrm{d}t} \geq 0 \ . \tag{3.50}$$

This corresponds to a *positive* operator $\hat{G}_{\text{ret}}$ (as can most easily be shown by means of the *linear* measure of entropy).

A phenomenological probability-conserving Markovian master equation for a system with 'macroscopic states' described by a (set of) 'relevant' variable(s) α, that is, $\rho_{\text{rel}}(t) \equiv \rho(\alpha, t)$ (see also Sects. 3.3 and 3.4) can be written in the general form

$$\frac{\partial \rho(\alpha, t)}{\partial t} = \int \left[w(\alpha, \alpha')\rho(\alpha', t) - w(\alpha', \alpha)\rho(\alpha, t) \right] \mathrm{d}\alpha' \ . \tag{3.51}$$

The transition rates $w(\alpha, \alpha')$ here define the phenomenological operator $\hat{G}_{\text{ret}}$ by means of its integral kernel $\hat{G}_{\text{ret}}(\alpha, \alpha') = -w(\alpha, \alpha') + \delta(\alpha, \alpha') \int w(\alpha, \alpha'')\mathrm{d}\alpha''$. They often satisfy a generalized time inversion symmetry,

$$\frac{w(\alpha, \alpha')}{\sigma(\alpha)} = \frac{w(\alpha', \alpha)}{\sigma(\alpha')} \ , \tag{3.52}$$

where $\sigma(\alpha)$ may represent the density of the ('irrelevant') microscopic states with respect to the variable α – that is, $\sigma := \mathrm{d}n/\mathrm{d}\alpha$, where n is the number of microscopic states as a function of α. In this case one may again derive an H-theorem, in analogy to (3.12), for the *generalized H-functional*

$$H_{\text{gen}}[\rho(\alpha)] := \int \rho(\alpha) \ln \frac{\rho(\alpha)}{\sigma(\alpha)} d\alpha = \overline{\ln p} \, . \qquad (3.53)$$

The final form on the RHS is appropriate, since the mean probability $p(\alpha)$ for individual microscopic states and for given $\rho(\alpha)$ is then $p(\alpha) = \rho(\alpha)/\sigma(\alpha)$. The entropy defined by $-kH_{\text{gen}}$ is also known as the *relative entropy* of $\rho(\alpha)$ with respect to the measure $\sigma(\alpha)$. The latter is often introduced *ad hoc* as part of a phenomenological description.

Under the approximation $w(\alpha', \alpha) = f(\alpha)\delta'(\alpha - \alpha')$ one now obtains the deterministic 'drift' limit of the master equation (3.51) – usually representing the first term of (3.44). It defines the first order of the *Kramers–Moyal expansion* for $w(\alpha, \alpha')$, equivalent to an expansion of $\rho(\alpha', t)$ in terms of powers of $\alpha' - \alpha$ at $\alpha' = \alpha$. The second order, $w(\alpha', \alpha) = f(\alpha)\delta'(\alpha - \alpha') + g(\alpha)\delta''(\alpha - \alpha')$, leads to the *Fokker–Planck equation* as the lowest non-trivial approximation that leads to an irreversible equation (see de Groot and Mazur 1962, Röpke 1987). In this respect, it is analogous to the LAD equation as the lowest non-trivial order in the Taylor expansion of the Caldirola equation (2.31). A master equation is generally equivalent to a (stochastic) *Langevin equation* for *individual* macroscopic trajectories $\alpha(t)$ which may form a dynamical ensemble represented by $\rho(\alpha, t)$.

In contrast to the Liouville equation (3.26), the master equation (3.48) or (3.36) cannot be unitary with respect to the inner product for probability distributions defined above (3.27). While total probability must be conserved by these equations, that of the individual trajectories *cannot* (see also Sect. 3.4). Information-reducing master equations describe an indeterministic evolution, which in general only determines an ever-increasing ensemble of different *potential* successors for each macroscopic state (such as a point in α-space).[7] As discussed above, this macroscopic indeterminism is compatible with microscopic determinism if that information which is transformed from relevant

---

[7] The frequently used picture of a 'fork' in configuration space, characterizing a dynamical indeterminism, may be misleading, since it seems to imply unique predecessors. This would be wrong, as can be recognized, for example, in an equilibrium situation. In the case of a stochastic dynamical law that is defined on a finite set of states, a state must in general also have different possible predecessors, corresponding to an inverse fork. Inverse forks by themselves would represent a pure forward determinism (a 'semigroup', that may describe *attractors*). All these structures are meant to characterize the *dynamical law*. They are neither properties of the (f)actual history (which is assumed to evolve along a *definite* trajectory regardless of the nature of the dynamical law), nor of an evolving ensemble that represents a specific state of knowledge.

However, only dynamically unique predecessors may give rise to recordable histories (consisting of 'facts' that are redundantly documented). The *historical nature* of our world is thus based on a uniquely determined or even overdetermined macroscopic past – see also footnote 1 of Chap. 2, Fig. 3.8, and Sect. 3.5. A macroscopic history that was completely determined from its macroscopic *past* would be in conflict with the notion of an (apparent) free will.

to irrelevant in the course of time no longer has any relevant (macroscopic) effects for all future times of interest. The validity of this assumption depends on the dynamics and on the specific initial conditions (3.46).

Time-reversed ('anti-causal') effects could only be derived from an appropriate *final* condition by applying the corresponding approximations to (3.44) for $t < t_0$. It is an *empirical fact* that such a condition, analogous to $A_{out}^{\mu} = 0$ in electrodynamics, does not describe our observed Universe. An exact boundary condition $\rho_{irrel}(t_0) = 0$ at some *accessible time* t_0 would for similar statistical reasons lead to a non-*decreasing* entropy for $t > t_0$, but to non-*increasing* entropy for $t < t_0$, hence to an entropy minimum at $t = t_0$ unless $S(t_0) = S_{max}$.

While the (statistically probable) assumption (3.46) led to the master equation (3.48), it would *not* necessarily characterize an arrow of time. Without an improbable initial condition $\rho_{rel}(t_i)$, the approximate validity of the equality sign in (3.50) would be overwhelmingly probable. Retarded action-at-a-distance electrodynamics would be trivial, too (and equivalent to its advanced counterpart) if *all* sources were already in thermal motion (such as the sources forming absorbers). It is the low entropy initial condition for ρ_{rel} which is responsible for the dynamical formation of that 'irrelevant' information which would be highly relevant for correctly calculating $\rho_{rel}(t)$ *backwards* in time.

The main conclusions derived in this and the previous section can thus be summed up as follows:

1. The *ensemble entropy* S_Γ does not represent physical entropy, since (a) it would be minus infinity for a real physical state (one or $N!$ points in phase space), (b) it is otherwise not additive for composite systems (in particular, it is not an integral over an entropy density), and (c) it remains constant under deterministic dynamics (in contrast to the Second Law). For indeterministic dynamical laws, it would have to increase, starting from its given value, in *both* directions of time (except when already at its maximum value). This demonstrates that ensemble entropy is *not* a physical quantity (see also Kac 1959).

2. Coarse-grained (or 'relevant') entropy, when defined as a function of the deterministically evolving microscopic state that is assumed to represent reality, would *most probably* fluctuate in time close to its maximum value. However, it may increase for a very long time – far exceeding the present age of the Universe if this had begun in an appropriate state of extremely low entropy (see Sect. 5.3). While a Zwanzig projection (describing generalized coarse-graining) can be arbitrarily *chosen for convenience* in order to derive an appropriate master equation, the cosmic initial condition must be *specified* as a condition characterizing the real Universe.

3. Only a relevance concept that includes locality is able to describe entropy as an extensive quantity.

4. Any coarse-grained entropy could be *forced* never to decrease by an appropriate modification of the corresponding *ensemble* dynamics – as in (3.36).
This may represent either new physics or an approximation to the situation described in the second part of item 2 (where the second possibility
is assumed to apply).

General Literature: Jancel 1963, Balian 1991.

3.3 Thermodynamics and Information

3.3.1 Thermodynamics Based on Information

As explained in the previous sections, Gibbs' probability densities or ensembles ρ_Γ represent incomplete *information* about the real state, which would in
classical mechanics be described by a singular point in phase space. Similarly,
Zwanzig's projection operators $\hat{P}$ (defining a generalized coarse-graining) were
justified by the incomplete observability of macroscopic systems. The entropy
and other parameters characterizing these ensembles, such as a temperature,
therefore appear fundamentally observer-related (objectively unmotivated).
While Gibbs' ensembles refer in principle to *actual* knowledge, Boltzmann's
distributions may be based on an objectivized limitation of knowledge, characterizing a certain class of potential observers, such as those able to recognize
only the mean particle density ρ_μ for a gas.[8] For similar reasons, the coarse-
graining $\hat{P}$ is kept fixed as a reference system, and not comoving according to the dynamics. The concept of information appears here extraphysical,
although observers or other carriers of information have to be regarded as
physical (in particular thermodynamical) systems, too (see Sect. 3.3.2).

Jaynes (1957) generalized Gibbs' statistical methods by rigorously applying Shannon's (1948) information concept. Shannon's formal measure of information for a probability distribution $\{p_i\}$ on a set of elements characterized
by the index i,

$$I := \sum_i p_i \ln p_i \leq 0 , \tag{3.54}$$

is evidently defined in analogy to Boltzmann's H, and therefore also called
negentropy. However, as a measure of information, it corresponds more closely
to Gibbs' extension in phase η. This measure is often normalized *relative*
to its value for minimum information, $p_i = p_i^{(0)}$, where $p_i^{(0)} = 1/N$ if $i = 1, \ldots, N$, unless different statistical weights for the 'elements' i arise from a
more fundamental level of description – cf. (3.53):

---

[8] The term 'objectivized' presumes the basically subjective (observer-related) status of what is to be objectivized. In contrast, the term 'objective' is in physics
often used synonymously with the term 'real', and then means the assumed or
conceivable existence of an object or its state regardless of its observation.

$$I_{\rm rel} = I(p_i|p_i^{(0)}) := \sum_i p_i \ln(p_i/p_i^{(0)}) = \ln N + \sum_i p_i \ln p_i \geq 0 \ . \qquad (3.55)$$

This renormalized measure of information may remain finite even when I diverges in the limit $N \to \infty$. Under an appropriate modification it can then also be applied to a continuum.

Jaynes thus based his approach on the idea that the microscopic state of a macroscopic system can never be completely *known*. Instead, a small though varying number of macroscopic variables, which are functions of the microscopic state, $\alpha(p, q)$, are approximately 'given'. Therefore, he introduced specific representative ensembles, $\rho_\alpha(p, q)$ or $\rho_{\bar\alpha}(p, q)$, which are defined to possess minimal information about all other variables (maximal ensemble entropy $S_\Gamma[\rho]$) under the constraint of either fixed values α, or fixed *mean* values $\bar\alpha := \int \alpha(p, q)\rho(p, q)\mathrm{d}p\,\mathrm{d}q$. This entropy thus becomes a *function* of α or $\bar\alpha$, defined as $S(\bar\alpha) := S_\Gamma[\rho_{\bar\alpha}]$, for example. This generalization of Gibbs' approach has turned out to be useful in many applications, while the macroscopic variables α remain to be chosen *ad hoc*.

As mentioned already in Sect. 3.1.2, an entropy concept based on the actually available information would be in conflict with the usual interpretation of entropy as an observer-independent physical quantity that can be objectively measured. On the other hand, its dependence on a certain basis of information may be quite meaningful. For example, the numerical value of $S_\Gamma[\rho]$ depends in a reasonable way on whether or not ρ contains information about actual density fluctuations, or about the isotopic composition of a gas. The probability $p_{\rm fluct}(\alpha)$ for the occurrence of some quantity α in thermodynamical equilibrium was successfully calculated by Einstein in his theory of Brownian motion from the expression

$$p_{\rm fluct}(\alpha) = \frac{\exp\left[S(\alpha)/k\right]}{\exp\left\{S[\rho_{\rm can}]/k\right\}} \ , \qquad (3.56)$$

thus exploiting the interpretation of entropy as a measure of probability. The probability for other quantities to be found immediately after the observation of this fluctuation would then have to be calculated from the 'conditioned' ensemble ρ_α rather than from $\rho_{\rm can}$.

Similarly, a star cluster (that is, a collection of macroscopic objects) possesses meaningful temperature and entropy $S \neq 0$ from the *point of view* that the motion of the individual stars is regarded as 'microscopic'. The same statistical considerations as used for molecules then show that their velocity distribution must be Maxwellian. At the other extreme, one could (in classical physics) conceive of an external Lapacean demon as a *super-observer* of the individual molecules in a gas. Entropy would indeed depend here on the available or accessible information. Its objectivity in thermodynamics can then only be understood as representing a common basis of information shared by us human observers. This perspective must be caused by our specific situation as physical systems.

In order to consistently regard ensembles as representing *actual* information, one would have to take into account all physical processes which affect the information carrier rather than just those in the system itself. Such a definition would certainly be inappropriate for the concept of physical entropy. For example, thermodynamical entropy does not depend on whether or how accurately the temperature has been measured; it is simply understood as a function of temperature.

Let $\alpha(p,q)$ represent a set of such quantities that are assumed to be 'given', possibly up to certain uncertainties $\Delta\alpha$ – see the model used in (3.51). The Hamiltonian $H(p,q)$ is in general just one of them. Subsets of microscopic states p,q corresponding to values of α within intervals $\alpha_0 < \alpha(p,q) < \alpha_0 + \Delta\alpha$, define subvolumes of Γ-space. The widths $\Delta\alpha$ may be those of Jaynes' representative ensembles for given mean values $\bar{\alpha}$, since any finer resolution would regard fluctuations as being relevant, unless α were a constant of the motion. For a single parameter α, these volume elements can be written as $\Delta V_\alpha := (dV/d\alpha)\Delta\alpha$, with $V(\alpha_0) := \int_{\alpha(p,q)<\alpha_0} dp\,dq$. In N-particle phase space, the size of the interval $\Delta\alpha$ is often quite irrelevant, since contributions to the volume integral for a compact region $\alpha(p,q) < \alpha_0$ may be strongly concentrated just below the surface defined by the value α_0 because of the geometry of such high-dimensional spaces. The term $\ln\Delta\alpha$ can then be neglected under the logarithm, $\ln\Delta V_\alpha$, that defines the entropy $S(\alpha)$.

One may now define a new useful Zwanzig projection $\hat{P}_{\mathrm{macro}}$ by *averaging* over subsets defined by such volume elements ΔV_α:

$$\hat{P}_{\mathrm{macro}}\rho(p,q) := \frac{\Delta p_\alpha}{\Delta V_\alpha} \tag{3.57}$$

$$:= \frac{1}{\Delta V_\alpha}\int_{\Delta V_\alpha} \rho(p',q')\mathrm{d}p'\mathrm{d}q' , \quad \text{for} \quad p,q \in \Delta V_\alpha .$$

If discrete values α_i are defined for convenience by means of 'macroscopic steps' $\alpha_i + \Delta\alpha = \alpha_{i+1}$, the integral for $S_\Gamma[\hat{P}_{\mathrm{macro}}\rho]$ splits into two sums:

$$S_\Gamma[\hat{P}_{\mathrm{macro}}\rho] = -k\int \hat{P}_{\mathrm{macro}}\rho\ln(\hat{P}_{\mathrm{macro}}\rho)\mathrm{d}p\,\mathrm{d}q$$

$$= -k\sum_i \Delta V_{\alpha_i}\frac{\Delta p_{\alpha_i}}{\Delta V_{\alpha_i}}\ln\frac{\Delta p_{\alpha_i}}{\Delta V_{\alpha_i}}$$

$$= -k\sum_i \Delta p_{\alpha_i}\ln\Delta p_{\alpha_i} + \sum_i \Delta p_{\alpha_i}k\ln\Delta V_{\alpha_i} . \tag{3.58}$$

[Note the relation to the concept of relative information (3.53) or (3.55) – see also Schlögl 1966.] The first term in the last line describes the entropy corresponding to the lacking *macroscopic information* described by the probabilities Δp_{α_i}. The second term is the *mean physical entropy* with respect to this macroscopic ensemble. The physical entropy, $S(\alpha) := k\ln\Delta V_\alpha \approx S_\Gamma[\rho_\alpha]$, thus measures the size of Jaynes' representative ensembles ρ_α, or, in Planck's language, the *number of complexions*, that is, the number of microscopic states which may represent it. In the special case $\alpha(p, q) := H(p, q)$ one obtains the entropy of the canonical ensemble as a function of the mean energy. If $\Delta\alpha = \Delta E$ is chosen infinitesimal, one obtains the entropy of the microcanonical ensemble, relevant for thermodynamically closed systems.[9]

Although the first term on the RHS of (3.58) is usually much smaller than the second one, it is essential for a complete and consistent discussion of information processing and measurement (see Sect. 3.3.2). A simple example of such a partitioning of the ensemble entropy into physical entropy and entropy of lacking information is provided by the particle number in a *grand* canonical ensemble, $Z^{-1}\exp\left[-(H - \mu N)/kT\right]$. This particle number is assumed to be 'given' (although in general not known) once the vessel that was in equilibrium with a particle reservoir characterized by the chemical potential μ has been closed. Thereafter, the system is represented by a *canonical* ensemble with fixed particle number N, while the relative contribution of that part of the original ensemble entropy which has now become entropy of lacking information about the exact particle number N is of the order $\ln N/N$ (Casper and Freier 1973). This contribution to the entropy is often neglected by using the 'approximation' $N! \approx N^N$. The argument demonstrates, however, that this different choice of ensembles is dynamically justified (by their robustness), and that the difference between the number of permutations, $N!$, of a fixed number N of particles and the factor N^N arising from the grand canonical ensemble with *mean* particle number N is meaningful – see (4.21) and cf. footnote 2.

The concept of physical entropy, defined above, no longer depends on actual information, since the choice of 'macroscopic' subsets, characterized by functions of state $\alpha(p, q)$, is motivated by their dynamical stability. In general, variables α characterizing 'robust' subsets of phase space that are densely populated by a trajectory (in the sense of quasi-ergodicity) within reasonably short times are regarded as macroscopic quantities. This quasi-ergodicity depends on a 'measure of distance' in Γ-space that cannot be invariant under canonical transformations. The macroscopic variables α are instead assumed to vary slowly and controllably – even under the influence of normal perturbations, or during their observation. These robust quantities define approximate constants of the motion or adiabatically changing collective variables. Since this concept of robustness is based on quantitative aspects, it cannot usually be defined with mathematical rigor. For example, the positions and shapes of droplets that are formed in a condensation process, or even more so those of the walls of the vessel, are evidently robust properties, although they do not represent exact constants of the motion.

---

[9] The infinite renormalization which is required for the corresponding concept of an entropy *density* as a function of α is due to the fact that the entropy for a continuous quantity has no lower bound, so that the measure of information may grow beyond all limits – see the remarks following (3.55).

A microscopic trajectory $q(t)$ determines all macroscopic trajectories $\alpha(t)$ defined as functions of this state: $\alpha(t) := \alpha(p(t), q(t))$. As discussed in Sect. 3.1.2, the macroscopic dynamics $\alpha(t)$ is then in general not autonomous, since trajectories starting from the same $\alpha(t_0)$ may evolve into different $\alpha(t_1)$ – depending on the microscopic initial state $p(t_0), q(t_0)$. This *macroscopic indeterminism* is essential for fluctuations or certain phase transitions.

The determinism of a dynamical *model* (such as Laplacean mechanics) is defined by the *mathematical existence* of a unique mapping of appropriate initial (or final) states onto complete trajectories. This concept of determinism is independent of the availability of an (analytic or algorithmic) *procedure* for explicitly constructing these trajectories in terms of conventional coordinates ('integrability'). It is therefore also independent of any *practical* limitation to their computability, which forms the basis of Kolmogorov's (1954) entropy, and is often used in the definition of *chaos* (see Schuster 1984, or Hao-Bai-Lin 1987). In classical mechanics, the deterministic dynamical mapping of initial conditions onto trajectories is a consequence of Newton's equations under non-singular conditions (see Bricmont 1996 for his lucid criticism of the popular misuse of the concept of chaos in this connection).

Trajectories could in principle be described in terms of the constants of the motion. The latter could then be used as new coordinates or 'co-evolving grids' (see Appendix B of Zurek 1989). Such constants of the motion are often denied to exist, since they are not analytically related to conventional coordinates. However, this does *not* mean that they would not exist in any absolute sense. It was indeed one of the great lessons from the theory of relativity that physics and spacetime geometry ('reality') are independent of the choice of coordinates, while the ancient Greeks were not even able to overcome Zeno's paradox of Achilles and the tortoise by a transformation to more appropriate 'coordinates' of description. We should similarly be able to conceptually overcome all mathematical problems in the construction of canonical transformations, and instead rely on the assumption of a coordinate-free 'reality' (at least in classical mechanics).

These mathematical difficulties may nonetheless reflect the complex and non-trivial *physical* relation between the Universe and its 'observing parts'. Observers are evidently *not* in any simple way related to the constants of the motion – the reason why we feel 'time change'.[10] Some authors have related the problems of a universe that contains its observers (*physical* self-reference) to Gödel's undecidability theorems, which apply to logical systems that allow *formal* self-reference (see Wheeler 1979). However, one cannot argue that the existence or meaning of an observer-independent reality is excluded just because of the observers' limited capabilities. This insufficient argument has even been used as an explanation of 'quantum uncertainty' (Popper 1950, Born 1955, Brillouin 1962, Cassirer 1977, Prigogine 1980). There is a fundamental

---

[10] "Time goes, you say? Ah no! Alas, time stays, we go." (Austin Dobson – discovered in Gardner 1967.)

difference between the *impossibility of ever knowing* the precise classical state of the Universe and the *incompatibility of its existence with certain empirical facts*. While the former is often derived precisely by using (thus presuming) classical concepts as describing reality, the latter is a consequence of crucial experiments on which quantum theory is based.

3.3.2 Information Based on Thermodynamics

Macroscopic indeterminism, such as described by Einstein's fluctuations in (3.56), may give rise to a transient *decrease* in physical entropy $S(\alpha)$ in accordance with microscopic determinism. It requires the transformation of lacking irrelevant into lacking relevant information. The latter would *not* be lacking any more if the fluctuation were observed, or, similarly, after the *measurement* of a microscopic variable, as depicted by the first step of Fig. 3.5. In these cases, the physical realization of information by observers or other information carriers has to be properly taken into account.

As is well known since the discussion of *Maxwell's demon*, any change or use of information must be described physically, with all its thermodynamical consequences. Maxwell had assumed his demon to operate a microscopic sliding door between two compartments of a vessel in such a way that only fast molecules may enter the first compartment, while only slow ones are allowed to leave it. His actions must then lead to a temperature and pressure difference, thus admitting the construction of a perpetuum mobile of the second kind.

The demon must here invest *its knowledge* about trajectories of individual molecules. However, Smoluchowski (1912) objected that a demon who acts physically would itself have to obey the Second Law: its operations must be described (thermo-)dynamically. In phenomenological terms, any lowering of the entropy of the gas must at least be balanced by a corresponding increase of the demon's entropy. If the demon were assumed to be a finite and thermodynamically closed system, its increasing Brownian motion would then ultimately prevent it from acting properly (by letting its 'hands tremble', or as a result of its deteriorating information about the molecules).

Szilard (1929) derived a fundamental information-theoretical consequence from this situation. By exploiting the idea of Maxwell's demon, he concluded that an 'intelligent being' must use up an amount of information of measure

$$\Delta I = \frac{\Delta S}{k} , \tag{3.59}$$

in order to *lower* the entropy of some system by ΔS. This equivalence would also be compatible with the ensemble interpretation of entropy, or Einstein's probabilities (3.56).

Szilard's main argument used a model 'gas' consisting of a single molecule in a vessel of volume V. Statistical aspects are introduced by means of many collisions of the molecule with the walls, leading to thermal equilibration

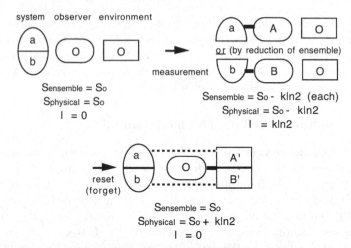

Fig. 3.5. Entropy relative to the state of information during a classical measurement. In the first step in the figure, the state of the observer changes depending on that of the system. The second step represents the subsequent resetting of the 'observer' or device (Bennett 1973), required if the process is to be exactly repeated for a second measurement. Areas represent sets of microscopic states of the subsystems (while those of uncorrelated combined systems would be represented by their direct products). The lower case letters a and b characterize the property to be measured; and 0, A and B the corresponding 'memory states' of the observer, while A' and B' are their respective effects in the thermal environment, required for a deterministic reset. The 'physical entropy' (*defined* to add for subsystems) measures the phase space of all microscopic degrees of freedom, including the property to be measured. Because of this presumed additivity, the physical entropy neglects statistical correlations (*dashed lines*, which indicate *sums* of direct products of sets) as being 'irrelevant' in the future – hence $S_{\text{physical}} \geq S_{\text{ensemble}}$. I is the amount of information held by the observer. S_0 is at least $k \ln 2$ in this simple case of two equally probable values a and b. (From Chap. 2 of Joos et al. 2003)

between the molecule's average motion and a surrounding heat bath (see Fig. 3.6). A piston is then inserted sideways (without using energy) in order to separate two partial volumes V_1 and V_2. This partition of the volume is *robust* in the sense of Sect. 3.3.1. According to (3.58), this procedure transforms part of the (physical) entropy of the 'gas' into entropy of *lacking information*. If the experimenter knows (only) in which partial volume $i = 1,2$ the molecule resides, corresponding to a Shannon measure $\Delta I_i = \ln[(V_1 + V_2)/V_i]$, he is able to retrieve the mechanical energy

$$\Delta A_i = \int_{V_i}^{V_1+V_2} p\mathrm{d}V = \int_{V_i}^{V_1+V_2} \frac{kT}{V}\mathrm{d}V = -kT \ln \frac{V_i}{V_1 + V_2} \tag{3.60}$$

by moving the piston into the empty volume, and slowly raising a weight, for example. The molecule's mean kinetic energy may thereby remain constant by

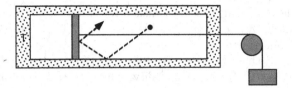

Fig. 3.6. Szilard's *Gedanken* engine completely transforms thermal energy into mechanical energy by using information

the reversible extraction of heat from the external reservoir with temperature T. This process lowers the entropy of the reservoir by

$$\Delta S_i = -\frac{\Delta A_i}{T} = k \ln \frac{V_i}{V_1 + V_2} = -k\Delta I_i \,, \tag{3.61}$$

in accordance with (3.59).

According to Smoluchowski, one could avoid referring to knowledge or information by using a 'mechanical rectifier' (such as a ratchet) that causes the piston to move in the appropriate direction. This rectifier would ultimately have to perform thermal motion large enough to make it useless, corresponding to the demon's trembling hands (see also Feynman, Leighton and Sands 1963, Vol. I, p. 46-1). So one has to conclude that utilizing knowledge for making decisions (for example in the brain) is equivalent to the operation of a rectifier. It is here essential that the rectifier cannot be reset to its initial state without getting rid of entropy – usually in the form of heat (Bennett 1987). For this reason the mechanism cannot work reversibly in a closed system.

Brillouin (1962), when elaborating on ideas originally presented by Gabor in lectures given in 1952 (see Gabor 1964), emphasized that Szilard's 'intelligent being' has to *acquire* information. Since this process must also be compatible with the Second Law, Brillouin postulated his *negentropy principle*

$$\Delta S' - k\Delta I \geq 0 \,, \tag{3.62}$$

which meant that any information gain ΔI has to be accompanied by some process of dissipation that leads to a *production* of thermodynamical entropy $\Delta S'$ *in the information medium* (usually light). He thereby referred to the latter's quantum aspect (photons), which limits its information capacity. Because of the minimum information required according to Szilard, the construction of a perpetuum mobile of the second kind would then again be excluded. However, because of the above example of a directly coupled mechanical rectifier, no explicit reference to an information medium seems to be required. According to Bennett (1973), it is the increase in physical entropy by the reset mechanism in Fig. 3.5, $\Delta S_{\text{phys}} = k \ln 2$ if $V_1 = V_2$, that compensates its decrease in (3.61).

All non-phenomenological arguments are based here on two assumptions: (1) Global determinism, which requires that an ensemble of N different states (or N ensembles of equal measure) must have N different successors, which have to be counted by the total ensemble entropy. Different states may evolve into the *same* final state only by means of an appropriate interaction with their environment, that transfers this difference to the latter (for example, in the form of heat). (2) Intuitive causality, which asserts that uncontrollable 'perturbations' by the environment can only *enlarge* the ensemble. It gives rise to *in*equalities such as (3.62) rather than equations. If thermodynamical concepts apply, a transfer of entropy ΔS must be accompanied by a transfer of energy according to $\Delta Q = T\Delta S$. This relation has also led to the interpretation of entropy as a measure of *degradation of energy*.

The equivalence of information and negative entropy suggests that any (tautological) information *processing* (for example in a computer) can in principle be performed reversibly. However, arithmetic operations are often *logically irreversible* in the sense that two factors cannot be recovered from their product. (In a mechanical computer this operation may indeed require friction.) This led to the conjecture that a minimum amount of entropy $k \ln 2$ has to be produced for each bit of information in each elementary calculational step (Landauer 1961). It was refuted by Bennett (1973 – see also Bennett and Landauer 1985). However, in their discussion the logically lost information ('garbage bits') – even if randomized – is still regarded as macroscopic or 'relevant' in the thermodynamical sense. For this reason, the entropy creation is deferred to the reset or clearing of the memory, which is required for the computer to perform its calculational steps more than once (see the second step of Fig. 3.5). These considerations will lead to quite novel consequences for quantum computers (see Sect. 4.3.3).

All these arguments support the interpretation that information has to be *physically* realized (and therefore to be compatible with the laws of thermodynamics), rather than representing an extraphysical concept that has to be independently postulated for a statistical foundation of thermodynamics. On the other hand, *mathematical theorems* do *not* represent information (as, for example, assumed by Landauer 1996). Logic deals exclusively with tautologies ('analytical judgements') – as complicated as they may appear to our limited intelligence.

General Literature: Denbigh and Denbigh (1985), Bennett (1987), Leff and Rex (1990).

3.4 Semigroups and the Emergence of Order

In physical systems, 'ordered' states are characterized by low entropy. Order may appear in the form of simple structures (such as regular lattices) or complex ones (organisms). For example, the rectifier discussed in the previous section as replacing Maxwell's demon must display ordered *dynamical* behavior. The emergence of order from disorder in Nature, also called *self-organization of matter*, may appear to contradict the Second Law with its general trend towards disorder and *chaos*. This has often been misunderstood as a 'discrepancy between Clausius and Darwin'. However, the fundamental phenomenological equation (3.1) allows entropy to decrease *locally*. A negative first term would allow physical entropy to flow into the environment. If this environment is not in complete thermal equilibrium, and characterized by at least two different temperatures, T_1 and T_2, a local loss of entropy, $dS_{ext} = dQ_1/T_1 + dQ_2/T_2 < 0$, would not even require any net loss of heat, $dQ_1 + dQ_2 < 0$. (Here differentials are always meant to refer to positive time increments dt.) This local decrease of entropy is thus *not* in conflict with its global increase according to the Second Law – see also Sect. 5.3.

In statistical terms, the number of states in a *dynamically representative ensemble* (see Sect. 3.1.2) may decrease locally in accordance with determinism *and* intuitive causality, provided the ensemble characterizing the state of the environment increases accordingly – precisely as during the 'reset' of a memory device, indicated in Fig. 3.5. In this Laplacean description, the outcome of evolution would be determined by the microscopic initial state of the whole Universe.

An important special case is a *steady state* of non-equilibrium, characterized by $dS = dS_{int} + dS_{ext} = 0$ in spite of non-vanishing entropy production, $dS_{int} > 0$ (Bertalanffi 1953). It may support ordered states as *dissipative structures*. The standard example, known as *Bénard's instability*, describes convective heat transfer through a thin horizontal layer of a liquid in the form of spatially ordered convection cells, which optimize the process of thermal equilibration between two reservoirs at different temperatures. In a finite universe, this stationary situation can only represent a transient local phenomenon. The emergence of structure is often connected with symmetry breaking (in particular of translational symmetry), related to a phase transition. In a deterministic description, an initial microscopic fluctuation would thereby become unstable and be amplified to a macroscopic scale. In quantum theory, it may also require an indeterministic collapse of the wave function (see Sect. 4.1.2).

For similar reasons, Boltzmann suggested that biological processes here on earth are facilitated by the temperature difference between the sun (with its 6000 K surface temperature) and the dark Universe (at 2.7 K, as we know today). At the distance of the earth, the solar radiation has an energy density much lower than that of a black body with the same spectrum (temperature). Since photon number is not conserved (in general not even a robust quantity), a canonical distribution $\exp(-H/kT)$ in the occupation number representation determines not only the spectral distribution as a function of temperature, but also the intensity (photon density). A gas with conserved particle number would instead allow one independently to choose the mean density – either by fixing the particle number by closing the vessel, or by fixing the chemical potential (in a grand canonical ensemble) by connecting the vessel to a particle reservoir. In contrast, a photon from the sun can be transformed very efficiently into many soft photons, which together possess much higher physical entropy.

Although order appears to be an objective property, an absolute concept of order that is not simply defined by means of phenomenological entropy is as elusive as an objective concept of information or relevance (see Denbigh 1981, p. 147, or Ford 1989). For reasons already mentioned in Sect. 3.3.1, the definition of order in terms of computability would depend on the choice of 'relevant coordinates'. For example, the obvious order observed in a crystal lattice is not invariant under general canonical transformations. How, then, may the order of an organism be conceptually distinguished from the 'chaotic' correlations arising from molecular collisions in a gas?

Many self-organizing systems include chemical reactions. They are phenomenologically described by irreversible rate equations, which define the dynamics of concentrations $X, Y, \dots$ These concentrations are 'macroscopic' variables, called α in Sects. 3.2 and 3.3.1. In statistical terms, rate equations can be derived from a generalized *Stoßzahlansatz* that includes rearrangement collisions between different kinds of molecules, which are usually assumed to be already in *thermal* equilibrium with one another. These rate equations are therefore special master equations (as derived in Sect. 3.2) for these 'relevant' degrees of freedom $X, Y, \dots$

Rate equations determine trajectories in the configuration space of concentrations.[11] For closed systems, these trajectories may eventually approach that point in their configuration space which describes equilibrium. Reversible determinism must come to an end at such *attractors* (see Fig. 3.7a), although this may require infinite time. A *mechanical* example of an attractor in the presence of friction is the phase space point characterized by $v := dx/dt = 0$ and $V(x) = V_{min}$. The corresponding equation of motion, $m\,dv/dt = -av - \nabla V$, neglects any stochastic response from the energy-absorbing microscopic degrees of freedom, which is in principle required by the *fluctuation–dissipation theorem*. Similar to the LAD equation of Sect. 2.3, this equation is, therefore, deterministic, even though it is asymmetric under time reversal.

Points in the space of macroscopic variables X, Y or x, v ('macroscopic states' α, in general) describe the physical states incompletely. They represent large subspaces of the complete Γ-space (that may realistically even have to include the environment). Volume elements of the same size in macroscopic

---

[11] As the rate equations are of first order in time, this macroscopic configuration space is often called a phase space.

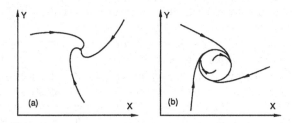

Fig. 3.7. Standard representation of an attractor (**a**) and a limit cycle (**b**) as examples of phenomenologically irreversible dynamics in the configuration space of macroscopic variables $\alpha \equiv X, Y$

'phase space' may correspond to very different ensemble measures. These volume elements are therefore in general not dynamically conserved. For example, the immediate vicinity of an equilibrium 'state' X_0, Y_0 – such as $v = 0$, $V(x) = V_{\min}$ in the mechanical example – covers almost the whole Γ-space of the completely described system (or some subspace that is defined by conserved quantities).

In the specific mechanical example with friction, the *modified* macroscopic phase space measure $\mathrm{d}x \, \mathrm{d}v/v$ nonetheless happens to be dynamically invariant. Time reversal is here compensated for by a transformation $v \to 1/v$ to restore a formal T-symmetry of macroscopic determinism (in formal analogy with the examples mentioned in the Introduction). This leads to a conserved *generalized H-functional* – cf. (3.53) and (3.55), viz.,

$$H_{\mathrm{gen}} := \int \rho(v, x) \ln \big[|v| \, \rho(v, x) \big] \mathrm{d}v \, \mathrm{d}x \, , \qquad (3.63)$$

which defines a 'reference density' $\rho_0 = |v|^{-1}$ as an effective equilibrium measure on this macroscopic phase space.

In the situation of a steady state non-equilibrium, macroscopic trajectories described by other effective irreversible equations of motion may approach certain *closed curves*, which do *not* correspond to maximum entropy. They are called *limit cycles*, and may represent dissipative structures, which represent order (see Fig. 3.7b and Glansdorff and Prigogine 1971).

Open systems are often described by means of *phenomenological semigroups*, defined as dynamical maps acting on ensembles in finite time steps. These maps can be understood as time-integrated Markov operators G_{ret}, and are thus applicable again only in the 'forward' direction of time (in contrast to the reversible group of time translations, valid for dynamically closed systems). Mathematically, ensembles may even be regarded as the fundamental kinematical objects of the theory, without any explicit definition of their *elements*, which would describe microscopic reality. 'Determinism' is then understood as a mere forward determinism for these formal ensembles. *Maps are* called irreversible if they form genuine semigroups, that is, if they cannot be uniquely inverted *as maps on ensembles*.

This irreversibility *of maps* does *not* correspond to a dynamical indeterminism for elementary states: it usually represents resets or attractors phenomenologically – that is, without explicitly taking into account microscopic degrees of freedom. In order to describe a reset, the master equation (3.48) has to be based on a *non-Hermitean* Zwanzig projector that 'creates' relevant information. In a globally deterministic context, its microscopic realization would then have to contain some way of getting rid of entropy (as discussed in Sect. 3.3.2).[12] As can be seen from the second step of Fig. 3.5, the reset transforms local information into nonlocal correlations (also depicted in Fig. 3.1). This transformation describes a *production of physical entropy*, while the ensemble entropy is conserved. The absence in Nature of correlations which would allow the inverse process and thus lead to a reduction of physical entropy is responsible for the irreversibility of the semigroup.

As these semigroups are defined to act on ensembles, regarded as abstract objects, their inversion does *not* in general represent a reversal of the microscopic dynamics ('time reversal'). For the same reason, their forward determinism is not equivalent to microscopic determinism. A dynamical map may not be invertible *as a map* even though the underlying dynamical transformation of microscopic states can be reversed.

Individual indeterminism and attractors are illustrated on a finite set of states in Fig. 3.8. An *asymmetric* dynamical indeterminism (b) is represented by diverging forks (see footnote 7), while an attractor is characterized by converging (or 'inverse') forks (c). An everywhere defined indeterminism must apply symmetrically (a). On a continuum of states, one would first have to define a measure, usually according to its invariance under the assumed *fundamental* deterministic dynamics of the completely described closed system. (This may represent a problem if determinism is to be given up fundamentally.) Semigroups are often studied on discrete state spaces, where measures of states are trivial. A popular example is the model of 'deterministic cellular automata' (see Kauffman 1991). Their merging trajectories (representing attractors) then replace the shrinking phase space in the continuum model with friction that led to the generalized H-functional (3.63).

Forward-deterministic dynamical maps are often defined by means of nonlinear transformations. A popular (though not very physical) one-dimensional toy model of a semigroup is the *Bernoulli shift*, defined by the mapping

---

[12] Therefore, mathematical physicists have proposed a new definition of entropy that would *always* allow the entropy of an open system to grow under a semigroup, even if its physical entropy decreased – see Mackey (1989). For example, the *relative entropy* – cf. (3.53) and (3.55)], defined for an open system with respect to a canonical distribution with temperature of an *external heat bath*, would increase even when the temperature of the system is *lowered* (as it is in the case of a cooler heat bath). Such a formal redefinition of entropy is certainly physically misleading, even though it may be useful for certain purposes.

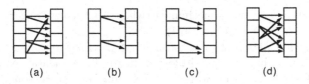

Fig. 3.8. Dynamical transformations of states on a discrete and finite 'phase space' consisting of only four states: (**a**) T-symmetric indeterminism (representing an incompletely determined Hamiltonian, for example); (**b**) asymmetric indeterminism, representing a law-like increase of ensemble entropy (cannot be defined everywhere on finite sets); (**c**) attractors (cannot be inverted as a map defined on all ensembles); (**d**) discrete caricature of a Frobenius–Perron map (see text). The symmetric indeterminism (**a**) would appear asymmetric – similar to (**b**) – when *applied to* a low-entropy initial ensemble (such as an individual state) in a given direction of time. It would then describe the usual increase of ensemble entropy by uncontrollable 'perturbations'. The distinction between (**b**) and (**c**) requires an absolute direction of time

$\alpha \to 2\alpha \,\mathrm{mod}\,1$ on the interval $(0, 1]$. (Its α-measure would be invariant under translations in α if chosen as 'fundamental dynamics'.) The dynamical increase of this 'phase space volume' element $d\alpha$ by multiplication by the factor 2 in this map could be uniquely inverted on the *infinite* continuum, although it represents an indeterminism in the sense of the measure. However, the second term, $\mathrm{mod}\,1$, characterizes a semigroup as in Fig. 3.8c. Both dynamical parts are combined here in order to form a *Frobenius–Perron map*, defined everywhere on the interval in spite of representing a semigroup (symbolically indicated for the discrete case of Fig. 3.8d). The forward indeterminism, obvious in the discrete case, is often overlooked on the continuum, where the topology-conserving stretching of 'phase space' α may appear deterministic without a measure. The 'topological time asymmetry' $(\mathrm{mod}\,1)$ contained in the Frobenius–Perron map may be phenomenologically useful, as it is able to describe the formation of macroscopic diversity. Realistic attractors must be of mixed type in order to comply with the fluctuation–dissipation theorem.

Many similar dynamical maps are discussed in the literature. They are (at most) of phenomenological value, and have little explanatory power from a fundamental statistical point of view. Their investigators often seem to regard the underlying individual microscopic reality as irrelevant. The ensembles being mapped dynamically are then treated as *real states* of physical objects. This must, of course, lead to confusion from a fundamental point of view. Statistical theories based on dynamical maps are occasionally even used for a 'minimal' interpretation of quantum mechanics (see Sect. 4.4). The misuse of purely formal ensembles as describing *physical* states is thereby reversed by identifying wave functions (that is, elementary quantum mechanical *states*) with ensembles. However, the conclusion that quantum phenomena cannot be explained in any such 'simple way' was already drawn by Bohr before the advent of matrix and wave mechanics (when his theory with Kramers and Slater had failed – see Jammer 1974).

The formation of structure is often related to a spontaneous symmetry breaking that may indeed have its origin in the fundamental *quantum indeterminism* (see Chap. 4 and Sect. 6.1). This may be the reason why the description of thermodynamical systems far from equilibrium (where structure may form) usually remains phenomenological (see Glansdorff and Prigogine 1971). The onset of structure may then be described by means of unstable fluctuations in certain quantities α, whose probabilities can be calculated from Einstein's formula (3.56). An instability would arise for them when the second derivative $\partial^2 S/\partial \alpha^2$ at a stationary point of $S(\alpha)$ becomes negative, for example by an adiabatic change in an external parameter. In this way, new *robust* quantities in the sense of Sect. 3.3.1 (see also Sect. 4.3.2) may emerge, while physical entropy is transformed into entropy of the corresponding *lacking information*, defined according to (3.58).

General Literature: Glansdorff and Prigogine 1971, Haken 1978, Cross and Hohenburg (1993).

3.5 Cosmic Probabilities and History

I shall close this chapter with a brief discussion of an objection against the probability interpretation of entropy when applied to the whole Universe and its evolution. It was first raised by Bronstein and Landau (1933), and later in a more explicit form by von Weizsäcker (1939) – see also Feynman (1965), but it may also be affected by some recent developments in cosmology.

The present state of the Universe does not only possess an entropy $S_{\alpha(\text{now})}$ that is much smaller than its equilibrium value S_{equil}; it also contains documents which strongly indicate that the entropy has always been increasing during the past, $dS_{\alpha(t)}/dt > 0$ for $t < t_{\text{today}}$. One may now compare the probability that these documents (including our private memories) have indeed formed in such a *historical* process with the probability for their formation in a mere chance fluctuation. In the former case one has

$$S_{\alpha(\text{yesterday})} < S_{\alpha(\text{now})} \ll S_{\text{equil}} . \qquad (3.64)$$

However, if Einstein's measure of probability in terms of entropy (3.56) were applicable to the Universe, the formation of its present state in a chance fluctuation – as improbable as it may appear – would be far more probable than a state with much lower entropy in the distant past. This objection evidently undermines Boltzmann's explanation of the thermodynamical arrow of time as arising from a grand fluctuation that occurred in an eternal universe (see also Sect. 5.3), since this fluctuation could be replaced by a far smaller one when its size is measured in terms of entropy.

This probability argument requires that the left inequality (3.64) is valid not only with respect to physical (local) entropy, but also for an appropriate ensemble entropy that takes into account all those non-local correlations which represent the convincing *consistency of documents*. Their existence in a historical universe is related to Lewis' 'overdetermination of the past' (see footnote 1 of Chap. 2).[13] While the improbability of the present solar system, for example, as having occurred in a chance fluctuation would be 'moderate' compared to that for a corresponding whole universe, the former would then have to contain consistent though unexplainable documents about the latter. David Hume's fundamental insight that we can never *predict* anything with certainty (not even that the sun will rise again) applies to the past as well – even if we did not question the general validity of the dynamical laws. Strictly speaking, we cannot be sure about the existence of *any facts* that we seem to remember. The reliability of memories and documents is in principle as doubtful as that of predictions; only the *subjective local present* cannot be questioned. Hence, even Kant's premise that we are making experience cannot be taken for granted. Not what has been observed, only our (perhaps deceiving) 'memory' that we are aware of now is beyond doubt. Saint Augustine concluded in a similar way in his *Confessiones* that the past and the future 'exist' only in the present – namely as memory and expectation 'in the soul'. This long-standing philosophical debate seems to be deeply affected (though not overcome) by thermodynamical and statistical considerations.

However, Saint Augustine's epistemologically rigorous concept of reality is obviously too restrictive for the construction of a 'world model', which must in principle always remain hypothetical (Poincaré 1902, Vaihinger 1911). The probabilistic objection raised above, even if formally correct, will thus hardly be accepted as demonstrating that causality is an illusion, based on an accident. Einstein's probabilities (3.56) for the occurrence of non-equilibrium states α, motivated by the statistical interpretation of entropy, can indeed be justified only for those macroscopic properties α which have a chance of occurring repeatedly within relevant times ('quasi-ergodically') on a generic trajectory – that is, for properties which are *not robust* on relevant timescales (hence not for stable macroscopic properties).

Physical cosmology can fortunately be derived from the more *economical* hypothesis of a universe of finite age. A homogeneous (structureless) low entropy initial state appears more acceptable in this sense than a complex state with a similarly low value of entropy. Probabilities for later states can then be calculated as *probabilities for histories* (products of successive conditioned probabilities). For example, the folding of protein chains is usually calculated along trajectories of monotonically increasing entropy (according to a master equation). Final configurations not accessible through such histories would thus be excluded even when possessing relatively large entropy. (Quantum

---

[13] States containing consistent (though possibly deceiving) documents were called *time-capsules* by Barbour (1994a) – see Sect. 6.2.2.

mechanically, there is always a non-vanishing but extremely small tunneling probability for their occurrence.) Most probable under the initial condition are those final states that are accessible through the most probable histories. This picture *explains* consistent documents. The thus conditioned probability for an observable world such as ours having evolved *somewhere* in the Universe would even *grow* with its size (in contrast to the global initial probability). This argument may lend support to many kinds of 'multiverses' (see Tegmark et al. 2006), which are reasonable conceptions when extrapolated from the observable universe by means of empirically founded laws or symmetries.

Whether the situation of a universe which contains scientists observing it can be regarded as probable in this sense, or whether additional 'weakly anthropic' selection criteria are required[14], has hardly ever been estimated in a reliable and unbiased way. Only at a tremendously later age of our universe could a state of maximum entropy be reached via improbable intermediate states or through quantum mechanical tunnelling (Dyson 1979), such that unconditioned probabilistic arguments apply. The cosmologically very early time that we are living at may thus remain the major improbable fact.

A 'plausible' low-entropy initial state of the Universe will be considered in Sect. 5.3. Its discussion requires quantum theory. Quantum indeterminism, whatever its correct interpretation (see Sect. 4.6), may even allow the assumption of a *unique* 'initial' state of the Universe (with a very small entropy capacity) – see Chap. 6. However, it may be worth noticing that the outcome of evolution (including ourselves) must already have been *contained as a possibility* in the huge configuration space that represents the fundamental kinematical concepts – regardless of all probability arguments.

---

[14] The weak anthropic principle states that we are encountering a *rare* local situation (such as a planet like Earth or a special universe in a multiverse), since we could not exist somewhere else, while the strong principle requires that the whole Universe or Multiverse must fulfill very specific conditions in order to allow our existence as observers. It has even been claimed to possess 'predictive power'. The border line between the weak and the strong principle is shifting in modern cosmology. (See Barrow and Tipler 1986, and Sect. 6.1.)

4

The Quantum Mechanical Arrow of Time

The dynamics of probability distributions on classical phase space, discussed under various aspects in Chap. 3, may be *formally* translated into quantum mechanics by means of the canonical quantization rules. Many authors of standard textbooks therefore maintain that the foundation of irreversibility in quantum mechanics is identical to that in classical physics. There could then only be quantitative differences arising from different spectral properties of the 'corresponding' Liouville operators. However, this approach to statistical quantum mechanics completely ignores the fundamental interpretational differences of concepts that formally correspond to one another (such as probability distributions and density operators – see Sect. 4.2). It therefore conceals essential aspects of quantum theory which may be important for irreversibility in general (recall the general discussion in the Introduction):

1. The quantum mechanical probability interpretation represents an *indeterminism* of controversial origin. Most physicists seem to regard it as an objective *dynamical* indeterminism (see Fig. 3.8), and some even as representing a fundamental arrow of time that would go beyond dynamics. Others have instead suggested that one may *explain* the unpredictability of quantum mechanical measurement results in terms of conventional statistical arguments, viz., by means of thermal fluctuations that are related to the amplification process which leads to macroscopic outcomes. If, however, this question is circumvented by interpreting the wave function as representing 'human knowledge as an intermediate level of reality' (Heisenberg 1956), this may exclude *any* possibility of a dynamical analysis, while Maxwell's demon, discussed in Sect. 3.3.2, would return through the quantum back door. Therefore, the foundation of irreversibility seems to be intimately related to the *interpretation* of quantum theory (see Sects. 4.3 and 4.6). In order to clarify this situation as far as possible, one first has to analyze the dynamical formalism as it is actually *used*, and thus empirically justified.

2. The quantum theory is *kinematically nonlocal*. For example, the generic many-particle wave function $\psi(r_1, r_2, \ldots, r_N)$, which represents a 'pure' quantum state, describes *quantum correlations* that are *not* due to incomplete information (even though they may *lead to* statistical correlations in measurements). Similarly, a state of quantum field theory is given by a wave functional of fields which are defined all over space. This 'entanglement' is a direct consequence of the superposition principle. In quantum theory, *the state of the whole does not define states of its parts*. This is in fundamental contrast to the completely determined many-particle state of classical mechanics: a point in phase space (that is, a definite state) remains a point when projected onto a subsystem. The kinematical indeterminacy of the parts in quantum theory describes a non-trivial 'wholeness' of Nature, which cannot, as in classical physics, be interpreted as a mere *dynamical* interconnectedness (that may lead to *statistical* correlations in an incomplete description). Quantum nonlocality is not just a 'spooky action at a distance' that would affect *hidden local states*. Moreover, the absence of well defined subsystem states has nothing to do with Heisenberg's uncertainty (or 'indeterminacy') relations, which signal the limited validity of *classical* concepts for describing physical states. They apply even when the true and deterministically evolving quantum states are *certain* ('pure').

Consequences of these basic differences between classical and quantum statistical physics will be discussed after their formal analogy has been set up in Sect. 4.1.

4.1 The Formal Analogy

4.1.1 Application of Quantization Rules

The *formal* transition from classical to quantum statistical mechanics can be based on the 'canonical quantization rules', which replace functions of state $a(p, q)$ by 'corresponding' operators $A = a(P, Q)$, and Poisson brackets between them by commutators. For example, the Liouville equation (3.26) transforms as

$$i\frac{\partial \rho_\Gamma}{\partial t} = i\{H, \rho_\Gamma\} =: \hat{L}\rho_\Gamma \quad \longrightarrow \quad i\frac{\partial \rho}{\partial t} = [H, \rho] =: \hat{L}\rho . \quad (4.1)$$

It is then called the *quantum Liouville* or *von Neumann equation*. The classical probability densities $\rho_\Gamma(p, q)$ are thus replaced by *density operators* ρ. The caret is here used to distinguish the new operators, which act on the quantum mechanical Hilbert space operators (such as density operators), from these Hilbert space operators themselves. In the formal analogy, the new 'superoperators' (as they are sometimes called) correspond to the operators that were defined in Sect. 3.1.2 as acting on probability densities.

The Hilbert space operators form a *new* Hilbert space if an inner product $\langle\langle\rho_1,\rho_2\rangle\rangle := \mathrm{Trace}\{\rho_1^\dagger\rho_2\}$ is defined for them in analogy to the inner product $\langle\rho_{\Gamma 1},\rho_{\Gamma 2}\rangle = \int \rho_{\Gamma 1}^*(p,q)\rho_{\Gamma 2}(p,q)\,dp\,dq$ for classical probability densities on Γ-space – see the text above (3.27).

Furthermore, all mean values $\bar{a}$ of functions of state $a(p,q)$, defined with respect to probability densities $\rho_\Gamma(p,q)$, are replaced by *expectation values* $\langle A\rangle$ of corresponding 'observables' A:

$$\bar{a} := \int a(p,q)\rho_\Gamma(p,q)\,dp\,dq \quad \longrightarrow \quad \langle A\rangle := \mathrm{Trace}\{A\rho\}\ . \qquad (4.2)$$

Since (4.2) implies

$$\overline{\ln\rho} \quad \longrightarrow \quad \langle\ln\rho\rangle = \mathrm{Trace}\{\rho\ln\rho\}\ , \qquad (4.3)$$

the quantum mechanical entropy functional corresponding to the ensemble entropy S_Γ becomes *von Neumann's entropy*,

$$S[\rho] := -k\mathrm{Trace}\{\rho\ln\rho\}\ . \qquad (4.4)$$

However, in spite of this formal analogy, a density operator can no longer be interpreted as representing an ensemble of states that would define ensembles of values for all functions of state (see below).

The dynamics of the statistical operators, defined by the right-hand equation of (4.1), is unitary, with a formal solution

$$\rho(t) = U(t)\rho(0)U^\dagger(t)\ , \qquad (4.5)$$

and $U(t) = \exp(-iHt)$ for time-independent Hamiltonians. This dynamical form warrants conservation of von Neumann entropy (4.4) under the von Neumann equation,

$$\mathrm{Trace}\{\rho(t)\ln\rho(t)\} = \mathrm{Trace}\{U(t)\rho(0)U^\dagger(t)U(t)\ln\rho(0)U^\dagger(t)\}$$
$$= \mathrm{Trace}\{\rho(0)\ln\rho(0)\}\ . \qquad (4.6)$$

Since *classical* determinism (the conservation of probabilities along individual trajectories) may also be described in the form of a unitary time-dependence of probability distributions [see (3.26)], the formal argument in (4.6) may also be applied to classical ensemble dynamics.

The square of the Hilbert space norm of a density operator,

$$\|\rho\|^2 := \langle\langle\rho,\rho\rangle\rangle = \mathrm{Trace}\{\rho^2\} = \langle\rho\rangle\ , \qquad (4.7)$$

defines a *linear measure* of negentropy (see footnote 3 of Chap. 3). It is also conserved under the unitary dynamics (4.1). The corresponding linear entropy is often defined as $S_{\mathrm{lin}} = \langle\langle(1-\rho)\rangle\rangle$ (such that $0 \leq S_{\mathrm{lin}} < 1$). In contrast to this linear entropy, which uses the Hilbert space norm of operators, the *probability norm*, Trace$\{\rho\} = \langle 1 \rangle = 1$, characterizes a Banach space of *trace class operators* (those with non-vanishing trace). It is preferentially used in open systems quantum mechanics (Sect. 4.4), since total probability must be conserved even under phenomenological stochastic equations of motion that describe an increase in ensemble entropy.

In further formal analogy to classical ensemble mechanics, any coarse-grained (or relevant) information measured by Trace$\{(\hat{P}\rho)\ln(\hat{P}\rho)\}$ is in general *not* conserved under a unitary transformation. The Zwanzig projection operators $\hat{P}$ are once again idempotent operators on the Hilbert space of density operators, with the additional properties Trace$\{\hat{P}\rho\} = 1$ and positive $\hat{P}\rho$ for all ρ – just as in Sect. 3.2.

Statistical operators (density operators) ρ may be *represented* by various ensembles of wave functions ψ_α with probabilities p_α in the form $\rho = \sum_\alpha |\psi_\alpha\rangle p_\alpha \langle\psi_\alpha|$ (see Sect. 4.2). In the diagonal form of ρ, where the eigenstates ψ_α form an orthonormal set, $\|\rho\|^2$ is given by the sum $\sum_\alpha p_\alpha^2$. Its conservation (or that of $\langle \ln\rho \rangle = \sum p_\alpha \ln p_\alpha$) thus reflects the *individual* conservation of these diagonal elements in the moving basis $\psi_\alpha(t)$ – in analogy to the conservation of a comoving phase space volume in deterministic classical mechanics.

Matrix elements of the density operator with respect to a *random* basis $\{\phi_n\}$,

$$\rho_{mn} = \sum_\alpha \langle\phi_m|\psi_\alpha\rangle p_\alpha \langle\psi_\alpha|\phi_n\rangle = \sum_\alpha c_{\alpha m} p_\alpha c_{\alpha n}^* \tag{4.8}$$

if $\psi_\alpha = \sum_m c_{\alpha m}\phi_m$, are in general small for $m \neq n$ because of random phases of the coefficients in the sum over α. Pauli (1928) referred to this *random phase approximation* when he neglected off-diagonal matrix elements while deriving his master equation (4.18) below. However, they may in general be small individually in spite of amounting to a significant effect as a whole. In contrast, the complete *neglect* of off-diagonal elements *in a certain basis*,

$$\hat{P}_{\text{diag}}\rho_{mn} := \rho_{mm}\delta_{mn} , \tag{4.9}$$

defines the most important Zwanzig projection of quantum statistical mechanics. It regards these off-diagonal elements (or any interference between the states of this basis) as 'irrelevant' for all practical purposes, although it does not assume them to vanish. So it has nothing to do with the usual diagonalization of Hermitean operators in their eigenrepresentation. The inequality

$$\text{Trace}\{\hat{P}_{\text{diag}}\rho\ln(\hat{P}_{\text{diag}}\rho)\} = \sum_n \rho_{nn}\ln\rho_{nn} \leq \text{Trace}\{\rho\ln\rho\} = \sum_{mn} \rho_{mn}(\ln\rho)_{nm}$$

$$\tag{4.10}$$

is called *Klein's lemma* – see (3.35). It is a consequence of the fact that $\hat{P}_{\text{diag}}$ is a genuine projection operator (see Sect. 3.2).

An obvious (weaker) generalization of (4.9) is

$$\hat{P}_{\text{semidiag}}\rho := \sum_n P_n\rho P_n , \tag{4.11}$$

where $\{P_n\}$ (no caret!), with $P_m P_n = P_m \delta_{mn}$, is a complete set of projection operators on mutually orthogonal subspaces of the Hilbert space of quantum states. In quantum field theory, projections on 'unitarily inequivalent' separable subspaces of Hilbert space, sometimes even regarded as 'distinct Hilbert spaces', are often chosen for this purpose. However, these decompositions of non-separable Hilbert spaces are no less arbitrary than any other $\hat{P}_{\text{semidiag}}$ (though often *useful* in the case of large numbers of effective degrees of freedom). If imposed axiomatically, the relevance concept (4.11) may represent a *superselection rule* (Wick, Wightman and Wigner 1952, Jauch 1968, Hepp 1972). This observation suggests that proposed superselection rules are similarly based on some *dynamical* robustness like the 'thermodynamically macroscopic' variables of Chap. 3 that are usually assumed as 'given' – a possibility that will be further investigated and confirmed in Sect. 4.3.

4.1.2 Master Equations and Quantum Indeterminism

The Hamiltonian of a quantum mechanical system is often written in the form $H = H_0 + H_1$ in order to derive a master equation in terms of a perturbation expansion with respect to H_1. However, the main purpose of this split Hamiltonian is to define a relevance concept of type (4.9) or (4.11) by means of the eigenbasis of H_0. It may then (but need not) be *further* used for a time-dependent perturbation expansion with respect to the off-diagonal elements of H in this representation.

The dynamics of the 'relevant' part $\hat{P}_{\text{diag}}\rho$ is the dynamics of the diagonal elements of ρ. According to (4.1) one has in any representation (now writing $\hat{P}_{\text{diag}} = \hat{P}$ for short)

$$i\frac{d\rho_{mm}}{dt} = \sum_n (H_{mn}\rho_{nm} - \rho_{mn}H_{nm})$$

$$\equiv \sum_{n(\neq m)} (H_{mn}\rho_{nm} - \rho_{mn}H_{nm}) \,\hat{=}\, \hat{P}\hat{L}(1-\hat{P})\rho . \qquad (4.12)$$

Since the diagonal matrix elements of ρ do not contribute to the RHS, the first term of Zwanzig's pre-master equation (3.44), representing $\hat{P}\hat{L}\hat{P}$, vanishes for this relevance concept. The terms remaining in (4.12) describe the coupling to the 'irrelevant' off-diagonal elements, and demonstrate that the diagonal elements are dynamically autonomous only in the trivial case (see footnote 6 of Chap. 3 regarding the quantum Zeno effect). Because of the formal analogy, the rest of Zwanzig's method can then be applied, provided the required approximations are valid. The propagator $\exp\left[-i(1-\hat{P})\hat{L}\tau\right]$, occurring in the operator $\hat{G}_{\text{ret}}$ of the Markovian approximation (3.48), defines here a closed but highly non-trivial dynamics of the off-diagonal elements of ρ_{mn}.

Pauli's master equation can now be obtained from (3.48) and (3.45) by using a perturbation expansion in terms of the off-diagonal elements of the

Hamiltonian for calculating $\hat{G}_{\text{ret}} = \int_0^T \hat{G}(\tau)\mathrm{d}\tau$. These off-diagonal elements are thus assumed to be small, although the master equation would become trivial if they vanished exactly (that is, for $H = H_0$). This last remark emphasizes the *dynamical* role of the relevance concept.

Now consider the last three factors of the RHS of the integral kernel (3.45) applied to ρ:

$$(1 - \hat{P})\hat{L}\hat{P}\rho = (1 - \hat{P})[H, \hat{P}\rho] \mathrel{\hat{=}} H_{mn}(\rho_{nn} - \rho_{mm}) \quad \text{with } m \neq n . \quad (4.13)$$

This expression depends only on the off-diagonal elements of H. The projection $1 - \hat{P}$ is ineffective, as $\hat{P}\hat{L}\hat{P} = 0$. Similarly, one has for the *first* three factors of the RHS of (3.45), when applied to any matrix X:

$$\hat{P}\hat{L}(1 - \hat{P})X \mathrel{\hat{=}} \sum_{k(\neq m)} (H_{mk}X_{km} - X_{mk}H_{km}) . \quad (4.14)$$

Hence, $\hat{G}_{\text{ret}}$ is of second and higher orders in the off-diagonal elements of H. When neglecting higher orders according to Pauli, one has to express the remaining propagator $\exp\left[-\mathrm{i}(1 - \hat{P})\hat{L}\tau\right]$ in (3.45) solely in terms of diagonal elements of H, $H_{mm} =: E_m^{(0)}$. This means

$$\mathrm{e}^{-\mathrm{i}(1-\hat{P})\hat{L}\tau}X \mathrel{\hat{=}} \mathrm{e}^{-\mathrm{i}\left(E_m^{(0)} - E_n^{(0)}\right)\tau}X_{mn} , \quad (4.15)$$

and one obtains

$$\hat{P}\hat{L}(1 - \hat{P})\mathrm{e}^{-\mathrm{i}(1-\hat{P})\hat{L}\tau}(1 - \hat{P})\hat{L}\hat{P}\rho \mathrel{\hat{=}} \quad (4.16)$$

$$\sum_n |H_{mn}|^2 2\cos\left[(E_m^{(0)} - E_n^{(0)})\tau\right](\rho_{mm} - \rho_{nn}) .$$

This result corresponds to a Born approximation in terms of the off-diagonal elements of the Hamiltonian. The time integral required to obtain $\hat{G}_{\text{ret}}$ according to (3.49) leads to the resonance factor

$$\int_0^T \cos\left[(E_m^{(0)} - E_n^{(0)})\tau\right]\mathrm{d}\tau = \frac{\sin\left[(E_m^{(0)} - E_n^{(0)})T\right]}{(E_m^{(0)} - E_n^{(0)})} , \quad (4.17)$$

familiar from time-dependent perturbation theory. In the limit $T \to \infty$, this quotient becomes a δ-function times π, and (3.48) can be written (Pauli 1928)

$$\frac{\mathrm{d}\rho_{mm}}{\mathrm{d}t} = 2\pi \sum_n |H_{mn}|^2 \delta\left(E_m^{(0)} - E_n^{(0)}\right)(\rho_{nn} - \rho_{mm}) =: \sum_n A_{mn}(\rho_{nn} - \rho_{mm}) . \quad (4.18)$$

This *Pauli equation* is similar to other master equations, such as (3.51), while the coefficients A_{mn}, defined on the RHS, are transition rates in analogy to Boltzmann's $w(\boldsymbol{p}_1\boldsymbol{p}_2, \boldsymbol{p}_1'\boldsymbol{p}_2')$ of Sect. 3.1.1. If H_1 contains only two-particle interactions, the sum over n may indeed be written as a sum over particle pairs. According to the above definition, the coefficients A_{mn} conserve energy and satisfy the symmetry under collision inversion, $A_{mn} = A_{nm}$ [see (3.7)]. Therefore, the Pauli equation conserves total probability, $\sum_n \mathrm{d}\rho_{nn}/\mathrm{d}t = 0$.

The explicit form of the Pauli equation (4.18) may be used to discuss its range of validity, which must be limited by the approximations used when deriving the general master equation (3.48). It depends here on the spectrum of the Hamiltonian, which is often discrete for quantum systems. Nonetheless, Poincaré recurrence times can be neglected *in practice* for macroscopic quantum systems. Their energy spectra are usually so dense that they do *not* lead to any observable differences compared to a continuous spectrum. Quantum systems may even exhibit 'classical chaos' (Habib, Shizume and Zurek 1998). On the other hand, even a continuous spectrum would not by itself justify an arrow of time (as is often claimed). The negligibility of recurrences for all times of interest – whether they exist in principle or not – applies in *both* directions of time. The physical importance of the difference between discrete and continuous spectra seems to be grossly overemphasized in mathematical foundations of irreversibility.

However, the energy δ-function occurring in (4.18) is meaningful only inside an integral over energy E, or, as an approximation, under a sum over m. Therefore, Pauli combined groups of states with almost equal energies to form 'cells' (subspaces) representing a coarse-graining in order to apply a random phase approximation in the corresponding sums (see also van Kampen 1954). Erich Joos (1984) was able to show that the off-diagonal elements ρ_{mn} between states from such macroscopically different subspaces disappear by interaction with the environment ('decoherence' – see Sect. 4.3). This *dynamical* argument justifies Pauli's conceptual cells and his random phase 'approximation'.

When applied to a single initial state with $\rho_{00}(0) = 1$, Pauli's equation (4.18) assumes the form of *Fermi's Golden Rule* in the Born approximation. Replacing the sum over initial states n in (4.18) by an energy integral and a sum over all remaining quantum numbers β, that is, $\sum_n \cdots \longrightarrow \sum_\beta \int \sigma_\beta(E) \ldots \mathrm{d}E$ with a partial density of states $\sigma_\beta(E)$, and similarly substituting $m \longrightarrow E', \alpha$ for the final states, one obtains for the energy-integrated diagonal elements of final states $\alpha \neq 0$, $\rho_{\alpha\alpha} := \int \rho_{E'\alpha,E'\alpha}\sigma_\alpha(E')\mathrm{d}E'$:

$$\frac{\mathrm{d}\rho_{\alpha\alpha}}{\mathrm{d}t} = 2\pi|H_{\alpha 0}(E)|^2\sigma_\alpha(E) \quad . \tag{4.19}$$

Here, $H_{\alpha 0}(E) := H_{\alpha E, 0E}$, while α represents a 'decay channel'.

Although this Golden Rule (4.19) can thus be derived as an approximation from the unitary dynamics (4.12), it is mainly used to calculate *probabilities* for decay and other non-unitary 'quantum events' – conventionally described by a collapse of the wave function – see Sect. 4.6. (*Coherent* exponential decay according to the Schrödinger equation will be discussed in Sect. 4.5.) In contrast, Boltzmann's probabilistic transition rates $w(\boldsymbol{p}_1\boldsymbol{p}_2, \boldsymbol{p}_1'\boldsymbol{p}_2')$ refer to *ensembles of individually deterministic* collision trajectories (distinguished by their impact parameters). This different interpretation is facilitated by the fact that the formal concept of a density operator is already based on a probability interpretation (see Sect. 4.2). Nobody has ever been able to construct a model that would consistently explain the wave function as representing an ensemble of 'hidden variables'. (Bohm's theory, that *presumes* Schrödinger's wave function, will be discussed in Sect. 4.6.) In particular, the entropy (4.4) does not contain any contribution that might represent the missing information corresponding to such an ensemble (as in Fig. 3.5 for classical measurements).

Pauli's equation does indeed resemble Born's original formulation of the probability interpretation (Born 1926). Born used it to describe 'quantum jumps' between Schrödinger's stationary eigenstates of Hamiltonians H_0 that characterize isolated microscopic systems (such as atoms).[1] In quantum field theory, a similar splitting of the Hamiltonian is used to define the *interaction picture*. The special role attributed to the eigenstates of H_0 as representing the 'real' physical states, dynamically connected by discrete jumps, was historically motivated by their correspondence with Bohr's discrete atomic electron orbits. Quantum jumps (or a 'collapse of the wave function') are, of course, incompatible with deterministic *trajectories in Hilbert space*, that is, with time-dependent wave functions evolving according to a Schrödinger equation. The system Hamiltonians H_0 are thus assumed not to contain any interaction that would be responsible for stochastic transitions. This early attempt to objectivize the probability interpretation (or the observables used therein) by a dynamical process is therefore based on an essential approximation. (Recall the trivial result obtained for the Pauli equation in the exact energy basis!)

The general structure of the Pauli equation is preserved even when the perturbation expansion in terms of the off-diagonal elements of H (in a certain basis) is not used. This improved equation is known as Van Hove's 'exact' master equation (Van Hove 1957). It represents the master equation for the Zwanzig projection (4.9) without any *further* approximation. In particular, if the chosen basis of relevance (the eigenbasis of H_0) is the independent particle basis, the matrix elements H_{mn} appearing in the Pauli equation have to be replaced by the elements of a T-matrix, usually defined as $T := (S-1)/2\pi\mathrm{i}$, where S is the exact two-particle scattering matrix. This procedure presumes the negligibility of simultaneous many-particle collisions (just as Boltzmann's *Stoßzahlansatz*). However, the adjective 'exact' for Van Hove's equation is misleading even for a dilute gas, as it refers only to the calculation of $\hat{G}_{\mathrm{ret}}$,

---

[1] While Born may not have been using his concepts quite consistently in these early days of quantum mechanics, in his third (here quoted) paper on the probability interpretation he discussed probabilities for jumps between stationary *wave functions* – not probabilities for the occurrence of classical properties (such as particle positions). In scattering or decay 'events' he referred to plane waves as stationary states, which he then *associated* with particle momenta according to de Broglie's relation. One year before the formulation of the uncertainty relations this was not recognized as being in conflict (in principle) with the position measurement at the detector.

but not to the derivation of the master equation (3.48) in its preferred basis of relevance. Similarly to the choice of subspaces in (4.11), Born's probability interpretation, when applied to *measurements*, depends on the choice of appropriate 'observables'.

In analogy to the classical H-theorem (3.10), one may again show that the entropy corresponding to the Zwanzig projection $\hat{P}_{\text{diag}}$ never decreases under the Pauli or Van Hove equation:

$$\frac{\mathrm{d}S[\hat{P}_{\text{diag}}\rho]}{\mathrm{d}t} = -k\frac{\mathrm{d}\left(\sum \rho_{mm}\ln\rho_{mm}\right)}{\mathrm{d}t} \geq 0 \ . \tag{4.20}$$

Evidently, this entropy depends crucially on the chosen basis for diagonalization, that is, on the specific concept of relevance used in this master equation.

Because of the formal analogy, the classical canonical distribution, $\rho_{\text{can}}(p,q) = Z^{-1}\exp\left[-H(p,q)/kT\right]$, now becomes a canonical density operator, $\rho_{\text{can}} = Z^{-1}\exp(-H/kT)$. It can be derived precisely as in (3.19) by maximizing the entropy $S[\rho]$ under the constraint of fixed mean energy and probability norm. The so-called 'new statistics' (Bose or Fermi statistics) in terms of apparent *particles* is obtained when evaluating this canonical density operator in terms of quantum states of free *fields* – conveniently in the *occupation number representation*. Only when expressed in terms of particle states does it appear as a new method for counting them. The success of quantum statistics is indeed one of the strongest arguments against *particles* (in their original sense of pointlike objects in space, distinguishable by their trajectories) as a fundamental kinematical concept.

This conclusion, that fields rather than particles have to be quantized even for fields that never appear classically (such as spinor fields – see Zeh 2003), is also supported by the absence of Gibbs' self-mixing entropy (see footnote 2 of Chap. 3). The empirically correct measure on phase space, $\mathrm{d}^{3N}p\,\mathrm{d}^{3N}q/h^{3N}N!$, may then be obtained, for example, in the partition function Z for a grand canonical ensemble, $p_{E,N}(\mu,T) = \exp\left[-(E-\mu N)/kT\right]$. If this expression is evaluated by means of the familiar textbook approximation in the occupation number representation $|\{n_{\boldsymbol{k}}\}\rangle$ for spatial wave modes (often incorrectly regarded as 'single-particle' *wave functions*) with wave numbers $\boldsymbol{k} = \boldsymbol{p}/\hbar$ on a large space volume V, one obtains for dilute gases – where $N = \sum_{\boldsymbol{k}} n_{\boldsymbol{k}}$ and $E = \sum_{\boldsymbol{k}} \varepsilon_{\boldsymbol{k}} n_{\boldsymbol{k}}$, with $\varepsilon_{\boldsymbol{k}} = p(\boldsymbol{k})^2/2m$ and $\varepsilon_{\boldsymbol{k}} - \mu \gg kT$:

$$\begin{aligned}
Z(\mu,T) &= \sum_{\{n_{\boldsymbol{k}}\}}\exp\left[-\sum_{\boldsymbol{k}}\frac{(\varepsilon_{\boldsymbol{k}}-\mu)n_{\boldsymbol{k}}}{kT}\right] \\
&\approx \sum_N \frac{V^N}{h^{3N}N!}\exp\left(\frac{N\mu}{kT}\right)\int\exp\left(-\sum_{i=1}^{N}\frac{p_i^2}{2mkT}\right)\mathrm{d}^{3N}p \\
&\approx \sum_N \left[\frac{V}{h^3N}\exp\left(\frac{\mu}{kT}\right)\int\exp\left(-\frac{p^2}{2mkT}\right)\mathrm{d}^3p\right]^N \ .
\end{aligned} \tag{4.21}$$

The factorials $N! \approx N^N$ in the denominator are here *required* (as already known to Planck in 1900) in order to compensate for the sum over all permutations of the N momenta p_i in this N-fold integral, since they all represent the *same* oscillator quantum states for the various wave modes. The latter are described by wave numbers k which formally correspond to momenta p. The density matrix, and therefore the partition function, now factorize in terms of wave modes k rather than in terms of particle numbers, while the factorials do not have to be introduced ad hoc (as done by Satyendra Nath Bose in order to justify his photon concept).

General Literature: Jancel 1963.

4.2 Ensembles Versus Entanglement

Quantum wholeness is analyzable.

In the previous section, we derived the von Neumann equation from the Liouville equation by using the formal quantization rules. The dynamics of the density matrix, obtained in this way, is unitary. Therefore, it conserves $S[\rho]$, while the Pauli (or Van Hove) equation, albeit apparently derived from the von Neumann equation *as an approximation*, may seem to be superior, as it is able to describe quantum indeterminism and an increase in ensemble entropy, in particular in quantum measurements.

The Liouville equation itself was obtained in Sect. 3.1.2 by applying Hamilton's (that is, Newton's) equations to ensembles that represent incomplete knowledge about classical states. Since quantization of the Hamiltonian dynamics of mechanical systems leads to the Schrödinger equation, one may as well first quantize and then consider ensembles of its solutions $\psi_\alpha(t)$ with corresponding probabilities p_α, now describing incomplete knowledge *about the wave function* (see Fig. 4.1). This procedure may offer deeper insight into the meaning of the density matrix than its formal foundation of Sect. 4.1.1.

According to this ensemble interpretation, probabilities p_α rather than the density matrix $\rho(q, q')$ correspond *conceptually* to the probability distribution $\rho_\Gamma(p, q)$. The meaning of the density matrix can only be appreciated when considering ensemble expectation values of observables A, that is, *mean values of expectation values* with respect to different wave functions ψ_α :

$$\langle A \rangle := \sum_\alpha p_\alpha \langle \psi_\alpha | A | \psi_\alpha \rangle = \text{Trace}\{A\rho\} = \sum_n a_n \langle \phi_n | \rho | \phi_n \rangle , \qquad (4.22)$$

with

$$A := \sum_n |\phi_n\rangle a_n \langle \phi_n| \quad \text{and} \quad \rho := \sum_\alpha |\psi_\alpha\rangle p_\alpha \langle \psi_\alpha| .$$

The symbol $\langle A \rangle$ denotes here a twofold mean: with respect to the ensemble of quantum states ψ_α with their probabilities p_α, *and* with respect to

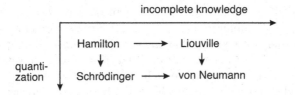

Fig. 4.1. Two routes from classical mechanics to the von Neumann equation

the quantum mechanical indeterminism of measurement results a_n with their probabilities $|\langle\phi_n|\psi_\alpha\rangle|^2$, valid for *given* quantum states ψ_α. In this way, the concept of a density matrix depends on the probability interpretation of the wave function – though *not* yet on any specific form in terms of ensembles[2] (see Sect. 4.6).

An ensemble interpretation of the density matrix according to $\rho = \sum_\alpha |\psi_\alpha\rangle p_\alpha \langle\psi_\alpha|$, used in (4.22), does not require the members ψ_α of the ensemble of wave functions to be mutually orthogonal; they may even form an overcomplete set. The ensemble can therefore not be recovered from the density matrix. Von Neumann's entropy (4.4) describes an ensemble entropy of the form $S[\rho] = -k\sum p_\alpha \ln p_\alpha$ only for the specific ensemble consisting of the orthonormal eigenstates of ρ.

Just as for classical statistical mechanics, the conservation of entropy reflects dynamical determinism (now for wave functions) – provided the Hilbert state norm is conserved, too. This requires not only determinism, but also the *unitarity* of the Schrödinger equation (not just that of the von Neumann equation). The reason is that the formal density matrix cannot distinguish between the norm and probability p_α of a wave function.

It should also be emphasized here that this formalism applies as well to wave functionals characterizing quantum field theory (that is, wave functions for a continuum of variables). 'Backward running' world lines in Feynman graphs are mere symbols for certain terms which appear in a relativistic perturbation expansion that is used for calculating the unitary propagation of wave functionals (general superpositions) with respect to an arbitrary but given time coordinate. These terms represent integrals over field modes (usually plane waves) – not over particle variables. Feynman's approach has turned out to be useful even beyond S-matrix theory, which is restricted to describing interactions between asymptotically free objects.

The mapping of *general ensembles* of wave functions onto those which diagonalize the density matrix is an information-reducing idempotent operation

---

[2] If the elements of the probability interpretation are themselves wave functions (as in Born's original formulation, mentioned in footnote 1, or as in collapse theories), the ensemble consisting of all possible outcomes of all conceivable measurements would be quite different from the initial ensemble (which may consist of one pure state, for example). Nonetheless, probabilities for all these outcomes are implicitly *postulated* by the phenomenological rules used in (4.22).

on these ensembles, similar to a Zwanzig projection. Nonetheless, one may rederive the von Neumann equation (4.1) from the ensemble interpretation under the further assumption that all wave functions defining the ensemble satisfy the same Schrödinger equation $i\partial\psi_\alpha/\partial t = H\psi_\alpha$. However, presuming the exact Hamiltonian to be 'given' is hardly consistent when regarding *states* as incompletely known. Even in classical physics, the precise Hamiltonian would depend on the (even less known) microscopic state of the environment (see Borel's argument in Sect. 3.1.2).

Instead of representing an ensemble of wave functions, the density matrix may also describe the local (or 'reduced') perspective of *entangled* quantum systems, which are generically of the form

$$\psi(x, y) = \sum_{m,n} d_{mn}\phi_m(x)\Phi_n(y) .\qquad(4.23)$$

For spatially separate subsystems, this entanglement defines *quantum nonlocality*. For example, it is responsible for the violation of Bell's inequality (Bell 1964) or its stronger variants (Greenberger, Horne, Shimony and Zeilinger 1990), and it explains so-called *quantum teleportation* in a way which demonstrates that nothing has to be teleported: it must rather be *prepared* in advance as a component of an entangled state (see Zeh 2005c or Timpson 2005).

All measurements performed on a subsystem – corresponding to the states $\phi(x)$, say – of an entangled system can be characterized by the expectation values for all its subsystem observables A_ϕ :

$$\langle A_\phi \rangle := \text{Trace}\{A_\phi\rho_{\text{total}}\} = \text{Trace}_\phi\{A_\phi\rho_\phi\} .\qquad(4.24)$$

Here, the 'reduced density matrix' ρ_ϕ is defined as a partial trace,

$$\rho_\phi := \text{Trace}_\Phi\{\rho_{\text{total}}\} .\qquad(4.25)$$

The total density matrix ρ_{total} may well be a pure state, $\rho_{\text{total}} := |\psi\rangle\langle\psi|$. The *new* density matrix ρ_ϕ would then be explicitly given in terms of the expansion coefficients d_{mn} of the total state (4.23) as

$$(\rho_\phi)_{mm'} := \langle\phi_m|\rho_\phi|\phi_{m'}\rangle = \sum_n d_{mn}d^*_{m'n} ,\qquad(4.26)$$

rather than in terms of probabilities p_α, which would instead lead to (4.8).

Both types of density matrices are Hermitean and positive by construction. They can therefore be diagonalized in the form $\rho_\phi = \sum_n |\tilde\phi_n\rangle p_n\langle\tilde\phi_n|$, with non-negative eigenvalues p_α, in their eigenbasis $\{\tilde\phi_n\}$. This diagonal form defines a formal (or apparent) ensemble of orthonormal states. Although the LHS of (4.26) is thus identical with a density matrix describing an ensemble of (orthogonal or other) states, it is evident from the RHS that it does *not* represent one. Therefore, the 'apparent ensemble' or 'improper mixture' (d'Espagnat 1966) must not be used in an attempt to *explain* the probability interpretation (4.24) on which it is based. The density matrix formalism is blind to the measurement problem (see below and Sect. 4.6).

For an entangled state such as (4.23), the eigenbases of the subsystem density matrices define the *Schmidt canonical form*,

$$\psi(x, y) = \sum_k \sqrt{p_k} \tilde{\phi}_k(x) \tilde{\Phi}_k(y) . \tag{4.27}$$

In contrast to the general representation (4.23) this is a single sum (Schmidt 1907, Schrödinger 1935). Phase factors for the coefficients $\sqrt{p_k}$ have here been absorbed into the phase-ambiguity in the definition of the orthonormal states $\tilde{\phi}_k$ or $\tilde{\Phi}_k$. For given subsystems, this representation (and hence its time dependence – see Kübler and Zeh 1973) is determined by the state $\psi(t)$ of the total system – except for accidental degeneracy of the p_k's.

The *neglect* of all correlations between two subsystems describes a specific loss of information, and so defines a new (nonlinear) Zwanzig projection,

$$\hat{P}_{\text{sep}} \rho := \rho_\phi \otimes \rho_\Phi . \tag{4.28}$$

A stronger Zwanzig projection of locality, $\hat{P}_{\text{local}} \rho = \prod_k \rho_{\Delta V_k}$, where the volume elements ΔV_k form a complete set of local subsystems, would lead to a density matrix that factorizes, as in (3.38), in terms of these volume elements. It is again required in order to obtain the approximate concept of an *entropy density* $s(\boldsymbol{r})$. In contrast to this local picture, indistinguishable particles *cannot* be used to define subsystems that might give rise to a 'substantial picture'. Therefore, the formal correlations between particles which describe symmetrization or antisymmetrization of the wave function does *not* represent any entanglement. These pseudo-correlations are merely an artifact from the use of classical particle concepts – see (4.21).

As a consequence of the nonlocality of quantum states, and in fundamental contrast to classical physics, the entropies $S[\hat{P}_{\text{sep}} \rho]$ or $S[\hat{P}_{\text{local}} \rho]$ of a completely defined (pure) quantum state are nontrivial: generically they do not vanish, since states of subsystems are *not defined* (rather than merely being unknown). *Apparent* ensembles, which are defined for them, may even be regarded as the *representative ensembles* used in statistical thermodynamics (see Chap. 3). However, one may now wonder (1) why microscopic systems are often found in pure states (such as eigenstates of their Hamiltonians H_0), and (2) why the macroscopic world is successfully described by means of *given* classical concepts rather than in terms of their superpositions.

A local concept of relevance that, in contrast to $\hat{P}_{\text{sep}}$, preserves all 'statistical' correlations (those based on incomplete information), while removing all quantum correlations (entanglement), may be defined by using the Schmidt canonical representation in the form

$$\hat{P}_{\text{classical}}(|\psi\rangle\langle\psi|) := \sum_k p_k |\tilde{\phi}_k\rangle\langle\tilde{\phi}_k| \otimes |\tilde{\Phi}_k\rangle\langle\tilde{\Phi}_k| . \tag{4.29}$$

Quantum correlations would here require a double sum over k and k' (in a non-Schmidt basis a sum over two *pairs* of indices). The RHS can be regarded as describing incomplete information about a presumed product state.

Quantum entanglement in bipartite systems has also been studied for 'mixed states' of the total system (Werner 1989, Peres 1996). Its consequences may be suppressed in such a mixture either by the presumed *averaging* over the ensemble of pure states that defines this mixed state, or by *tracing out the entangled environment* that gave rise to a reduced state for the total system. However, it would be quite inappropriate to define physical states as 'not' or 'less entangled' (as implicitly done by means of effective *measures of entanglement*) just because this entanglement cannot be confirmed by measurements in this situation.

A random *pure* state (4.23) would not lead to the 'statistical' result $\rho_{\text{irrel}} \approx 0$ – as used in (3.46) – for the relevance concept (4.28), since only the improbable factorizing states do not contain any correlations. (The reader may wish to skip the rest of this somewhat technical paragraph.) For example, the linear entropy according to (4.7), given by $S_{\text{lin}} = 1 - \sum_{kk'} |\rho_{kk'}|^2$ if normalized to vanish for a pure state, assumes its maximum in a Hilbert space of finite dimension D, $S_{\text{lin}}^{\max} = 1 - 1/D$, for the maximal mixture $\rho_{kk'} = \delta_{kk'}/D$. For *pure states* in the tensor product of two Hilbert spaces with dimensions M and N (hence with $D = MN$), one obtains for the *mean* linear subsystem entropy, defined below (4.7), in either subsystem (Lubkin 1978):

$$\bar{S}_{\text{lin}}^{\text{sub}} = 1 - \frac{M+N}{D+1} < 1 - \frac{1}{M} \quad \text{and} \quad < 1 - \frac{1}{N} . \qquad (4.30)$$

This entanglement entropy vanishes only for $M = 1$ or $N = 1$. Since the *linear information* $I_{\text{lin}} := 1 - S_{\text{lin}}$ factorizes for products of density matrices (rather than being additive as would its logarithmic measure), its value for the total system is given by $I_{\text{lin}}[\hat{P}_{\text{sep}}\rho_{\text{pure}}] = (I_{\text{lin}}^{\text{sub}})^2$, as the entropies of the subsystems must be equal for a pure total state – see (4.27). For $D \gg 1$ one has, according to (4.30),

$$\bar{I}_{\text{lin}}[\hat{P}_{\text{sep}}\rho_{\text{pure}}] \approx \frac{(M+N)^2}{MN} I_{\text{lin}}^{\min} ,$$

so that $\bar{I}_{\text{lin}}[\hat{P}_{\text{sep}}\rho_{\text{pure}}] \approx (M/N)I_{\text{lin}}^{\min}$ for $M \gg N$, with $I_{\text{lin}}^{\min} = 1 - S_{\text{lin}}^{\max} = 1/MN$ (characterizing maximal mixtures). While $\rho_{\text{irrel}} = (1 - \hat{P}_{\text{sep}})\rho$ vanishes in a random mixture $\sum_\alpha |\psi_\alpha\rangle p_\alpha \langle\psi_\alpha|$, the mean information for pure total states, $\sum_\alpha p_\alpha I_{\text{lin}}[(1 - \hat{P}_{\text{sep}})|\psi_\alpha\rangle\langle\psi_\alpha|]$, does not. Since the irrelevant information is measured by $I[\rho] - I[\hat{P}_{\text{sep}}\rho]$, rather than by $I[(1 - \hat{P}_{\text{sep}})\rho]$, the random pure state (with $S_{\text{lin}}[\rho] = S_{\text{lin}}^{\min} = 0$) must possess large *local* entropy $S_{\text{lin}}[\hat{P}\rho]$, and therefore has to carry its information predominantly in ρ_{irrel}.

Similar relations hold for the logarithmic entropy (Page 1993, Foong and Kanno 1994). It is impossible to reach $S^{\max}$ for $M \neq N$ for a pure total state. Because of $I_\phi = I_\Phi$, the *local information* about the larger system cannot be entirely transformed into correlations. However, every (small) subsystem of the completely described quantum universe would essentially possess maximum entropy for a random global state.

If the total wave function ψ evolves according to a Schrödinger equation, the reduced density matrix does *not* in general obey a von Neumann equation. While its exact dynamics can still be explicitly formulated (Kübler and Zeh 1973, Pearle 1979), it remains entangled with the rest of the total system. Indeed, the reduced density matrix ρ_ϕ, multiplied by the unit operator in Φ-subspace, defines a further (linear) Zwanzig projection,

$$\hat{P}_{\text{sub}}\rho_{\text{total}} := \rho_\phi \otimes 1_\Phi \ . \tag{4.31}$$

Phenomenological irreversible master equations for ρ_ϕ (instead of a von Neumann equation) are known as 'open systems' quantum dynamics (see Sect. 4.4). They are often derived by presuming an uncorrelated environmental heat bath (Favre and Martin 1968, Davies 1976). This assumption of lacking initial quantum correlations is similar to Boltzmann's chaos assumption. Master equations should therefore *explain* the canonical ensembles describing heat baths rather than presuming their existence. Open systems have also been described by means of path integrals (Feynman and Vernon 1963).

As already mentioned, both expectation values, (4.22) and (4.24), which were used to derive the concept of a density matrix, depend on the probability interpretation. Von Neumann (1932) introduced a model interaction in an attempt to describe ideal measurements (or 'measurements of the first kind') *dynamically*. It is defined by the unitary transformation $\phi_n\Phi_0 \xrightarrow{t} \phi_n\Phi_n$, where ϕ_n is an eigenstate of an observable $A = \sum_n |\phi_n\rangle a_n \langle\phi_n|$, while Φ_0 and Φ_n are the initial state of the apparatus and its final 'pointer positions', respectively. This dynamics describes a fork of causality in the classical *configuration space* on which the wave function is defined (see footnote 1 of Chap. 2). The observable A is thus *defined* by this interaction, whereby its eigenvalues a_n, characterize the 'pointer scale'. If the microscopic system is initially in one of the eigenstates of A, it does not change during such an ideal measurement, while the apparatus evolves into the corresponding pointer state Φ_n. In quantum optics, such measurements are also called 'quantum non-demolition measurements', since photons are usually absorbed when being measured. In the case $n = 1, 2$ and $\Phi_0 = \Phi_1$ they are identical to 'controlled-not gates' – much discussed in the theory of quantum computing.

However, for a general initial state, $\sum_n c_n\phi_n$, one now obtains for the same interaction, and for the same initial state Φ_0 of the apparatus,

$$\left(\sum_n c_n\phi_n\right)\Phi_0 \xrightarrow{t} \sum_n c_n\phi_n\Phi_n =: \psi_{\text{final}} \ . \tag{4.32}$$

If the pointer states Φ_n are mutually orthogonal, too, both sides of (4.32) are Schmidt-canonical. The RHS is now an *entangled* state, while an *ensemble* of different measurement results (that is, of states $\phi_n \Phi_n$ with probabilities $|c_n|^2$), would require the fork of causality to be replaced by a fork of indeterminism. (The formal 'plus' characterizing the superposition would have to become an 'or'.) This discrepancy represents the *quantum measurement problem*, that would be obscured in a phenomenological description by means of reduced density matrices for the subsystems only. The density matrices resulting from these two types of forks are identical, since there is no way of distinguishing these different situations operationally by a *local* measurement. As emphasized before, this argument does not *explain* the fork of indeterminism that lies at the heart of the probability interpretation.

This measurement problem prevails regardless of the complexity of the measurement device (that might give rise to thermodynamically irreversible behavior), and regardless of any perturbations caused by, and in the environment, since the states Φ may be assumed to describe this complexity completely, and even to include the whole 'rest of the Universe'. The popular argument that quantum mechanical indeterminism might, in analogy to the classical situation, be caused by thermal fluctuations that occur during a measurement process (see Sect. 3.3 or Peierls 1985, for example) is incompatible with universal unitarity. It would instead require the existence of an initial ensemble of microscopic states which in principle had to *determine* the outcome. However, the ensemble entropy of the RHS of (4.32) does not represent an ensemble that would allow the measurement to be interpreted as in Fig. 3.5.

If both systems in (4.32) are microscopic, the dynamics representing the fork of causality can even be reversed in practice (the measurement could be 'undone') in order to demonstrate that all components still exist. This reversal leads to observable consequences that may depend on all existing components, including their relative phases. This excludes the interpretation of (4.32) as a dynamical fork of indeterminism (a fork between mere possibilities), even though von Neumann's fork of causality is defined in terms of wave packets on a *classical configuration space*. Therefore, the transition from quantum to classical (Sect. 4.3) can be understood only if it explains why the fundamental arena for wave functions often *appears* as a space of classical configurations.

The interaction (4.32) is an example for the generic case of quantum mechanical subsystems which are not individually obeying unitary dynamics. Similar arguments would apply to the *ensemble* dynamics of systems with classical correlations (that is, if $\rho = \hat{P}_{\text{classical}}\rho$). In this case, the effective subsystem Hamiltonian H_ϕ, say, would depend on the state Φ_k of the complementary system by means of a partial expectation value,

$$H_\phi^{(k)}(t) := \langle \Phi_k(t)|H|\Phi_k(t)\rangle_{\text{Op}} , \qquad (4.33)$$

where H was defined to act on the tensor product of ϕ and Φ. There can be no Hamiltonian H_ϕ valid for the whole ensemble any more. This is equivalent to the induced Hamiltonians of interacting classical mechanical systems, which are given by

$$H_\phi(p_1 \ldots q_n, t) := H\big(p_1 \ldots q_n; p_{n+1}(t) \ldots q_N(t)\big) . \qquad (4.34)$$

Here, particle numbers $1, \ldots, n$ are meant to characterize the considered subsystem 'ϕ', while all others $(n+1, \ldots, N)$ represent the 'environment'. Each element of the ensemble would then satisfy another Hamiltonian or Schrödinger equation – in contrast to the assumptions leading to the Liouville or von Neumann equation. Nonetheless, for each element of an ensemble representing incomplete knowledge, the subsystem evolution would be *determined* in this classical case. Neglecting the statistical correlations dynamically by using $\hat{P}_{\text{sep}}$ in a master equation would amount to applying the whole resulting ensemble of sub-Hamiltonians (in the forward direction of time) to each individual element of the ensembles of states of the subsystems. However, only under the unstable assumption $\rho = \hat{P}_{\text{classical}}\rho$ (that is, without any entanglement) would the quantum mechanical situation simply be equivalent to the classical one of (4.34), or as in Sect. 3.1.2.

It should be kept in mind, therefore, that the local concepts of relevance, $\hat{P}_{\text{sep}}$, $\hat{P}_{\text{local}}$ and $\hat{P}_{\text{classical}}$, appear 'natural' only to our classical prejudice. In the unusual situation of controllable entanglement (as in EPR/Bell type experiments), quantum correlations may become relevant by means of the relocalization of superpositions even for local observers. *Dynamical* locality, as described by means of point interactions in field theory, merely warrants the dynamical consistency of these concepts of relevance, or gives rise to the approximate validity of autonomous master equations for $\hat{P}_{\text{local}}\rho$.

General Literature: d'Espagnat 1976, 1983.

4.3 Decoherence

> Novel ideas in science are at first completely neglected, then fiercely attacked, and finally regarded as well known.
>
> Konrad Lorenz

In Sect. 3.1 we saw how molecular collisions produce statistical correlations, which describe 'irrelevant' information. Although other relevance concepts may also be appropriate for describing irreversible phenomena, the formation of statistical correlations seems to be the most important one in classical description. In a gas, these correlations arise by means of a momentum transfer between molecules, eventually leading to a Maxwell distribution – the distribution of highest entropy for given mean energy if correlations are neglected.

If one specific 'molecule' happens to possess macroscopic mass (such as a bullet flying through the gas), its recoil may approximately be neglected in collisions with molecules – except for the resulting friction, whose importance depends on the density of the gas. The bullet may then remain in a non-equilibrium state of almost free motion for some time. On the other hand, collisions drastically affect the microscopic molecules. Although their states after scattering must strongly depend on the bullet's position at collision time, this dependence cannot be regarded as representing *information* about it if the molecular motions are already chaotic. In contrast, light scattered off the bullet *does* carry information, as we may easily confirm by using our eyes. The reason is that light interacts weakly or coherently with matter, and remains in a state far from equilibrium if absorption can be neglected (see Chap. 2).

The effect of an individual molecule or photon on a macroscopic object may thus be neglected in classical description, but this conclusion has to be radically revised in quantum mechanics. The quantum interaction can be described as an ideal (though uncontrollable) 'measurement' of the bullet's position and shape by the molecule in the sense of von Neumann. If the bullet were initially in a superposition of different positions, as one would have to expect for an object in a generic quantum state, this would lead to an entangled state as in (4.32). In this case, the initial superposition becomes *dislocalized* (it is at *no* place any more). This is called 'decoherence' if the dislocalization is irreversible in practice.[3] (Reversible dislocalization of a superposition – such as in a Stern-Gerlach device – may be regarded as 'virtual decoherence'.) It turns out that real decoherence is not only unavoidable for all macroscopic objects, but even the most abundant and most important irreversible process in Nature (Zeh 1970, 1971, 1973, Leggett 1980, Zurek 1981, 1982a, Joos and Zeh 1985).

In general, decoherence is not pure, but accompanied by a distortion of the system under consideration (recoil). For an environmental heat bath this would be required by the *fluctuation–dissipation theorem*, which leads to 'quantum Brownian motion' – a combination of decoherence, dissipation and fluctuation. However, the quantitative relation between these phenomena depends on actual parameters, such as temperature and mass ratios. Since fluctuation and dissipation may so become arbitrarily small, 'ideal' measurements by the environment are appropriate for studying 'pure decoherence' as a genuine quantum phenomenon. Chaotic molecules then contribute to decoherence just as ordered light. Evidently it is the physical effect on the environment that is essential – not any transfer of information. 'Quantum information' is here no more than a misleading renaming of entanglement.

Decoherence is also important for strongly interacting *microscopic* systems, such as individual molecules in a gas, although it is then far from being pure (recoil is essential). Instead of quasi-classical behavior, one now obtains quasi-stochastic dynamics – as successfully *used* in the *Stoßzahlansatz*. Interacting microsystems constituting solids can often be approximated by coupled harmonic oscillators (Caldeira and Leggett 1985). While solutions are then analytically available, they are also known to possess certain pathological properties. In particular, they are non-ergodic.

---

[3] The term decoherence has often been misused in the literature. See Sect. 3.4.3 of Joos et al. (2003) on 'True, False and Fake Decoherence'.

Applying the terminology used in the previous section, decoherence may be understood as the justification of a *specific* $\hat{P}_\text{semidiag}$ for a given subsystem by presuming the relevance of locality, as described by the corresponding $\hat{P}_\text{sub}$ – see (4.31). If this $\hat{P}_\text{semidiag}$ turns out to be dynamically valid under all normal circumstances, its eigenspaces characterize 'quasi-classical' properties or superselection rules (Zeh 1970, Zurek 1982a). Classical concepts *emerge* approximately in the form of apparent ensembles of narrow wave packets through unavoidable and practically irreversible interaction with the environment. They do not have to be presumed as an independent fundamental ingredient, required for an interpretation of the formalism (as done in the Copenhagen interpretation). From a pragmatic point of view, which does not distinguish between proper and improper mixtures, this would already be sufficient to solve the measurement problem. In a consistent description of reality in terms of wave functions, however, one must assume either a genuine collapse to be *triggered* by decoherence in some way, or appropriately redefine conscious observers within an Everett interpretation (see Sect. 4.6).

The interaction (4.32) was introduced by von Neumann to describe the controllable measurement of a microscopic system ϕ by an appropriate device (with 'pointer' states Φ_n). Its fact-like time asymmetry, leading from factorizing to entangled states, could be reversed with sufficient effort if both subsystems were microscopic ('recoherence' or 'erasure of measurement results'). For genuine quantum measurements, the pointer states Φ_n must be macroscopic. They are then 'measured' in turn by their uncontrollable environment, and thus become irreversibly quasi-classical. This explains why measurements which lead to macroscopic pointer positions cannot be undone.

It is this universality and unavoidability of entanglement with the natural environment that seems to have been overlooked for the first 50 years of quantum theory. All attempts to describe macroscopic objects quantum mechanically as being isolated, and therefore by means of a Schrödinger equation, were thus doomed to failure – even when including environment-induced dynamical terms that might describe a distortion. Decoherence is different, and extremely efficient, since it does not require an environment that *disturbs* the system. The distortion *of the environment by the system* affects the density matrix of the system, too, because of quantum nonlocality, but on a much shorter time scale than thermal relaxation or dissipation (Joos and Zeh 1985, Zurek 1986).

Some examples of decoherence will now be discussed in detail.

General Literature: Joos et al. 2003, Zeh 2005c, Schloßhauer 2006.

4.3.1 Trajectories

Imagine a two-slit interference experiment with bullets or small dust particles, described by quantum mechanics. Then not only their passage through the slits, but their whole path would be 'measured' by scattered molecules or photons. No interference fringes could ever be observed for such macroscopic objects – even if the resolution of the registration device were fine enough.

In this respect, macroscopic objects are similar to alpha 'particles' in a Wilson chamber, which interact strongly with gas molecules by means of their electric charge. For all these objects, their unavoidably arising entanglement with their environment leads to a reduced density matrix that can be represented by an ever-increasing ensemble of narrow wave packets following slightly stochastic trajectories (see also Mott 1929). This result is not restricted to the quantum description of motion in space: propagating wave packets in the configuration space of macroscopic variables may similarly explain their apparent 'histories'. For spatial motion the argument also demonstrates that the concept of an S-matrix does not apply to macroscopic objects, since it presumes asymptotically free states.

Several very instructive interference experiments have recently been performed with *mesoscopic molecules* that are in the transition region between isolated quantum and classical behavior. Various mechanisms of decoherence, including the emission of thermal radiation from internal molecular degrees of freedom, can be studied for them in detail (Arndt et al. 1999, Hornberger, Hackermüller and Arndt 2005).

For a continuous variable, such as position, decoherence competes with the dispersion of the wave packet that is reversibly described by the Schrödinger equation. Even the scattering rate of photons, atoms, or molecules off small dust particles in intergalactic space suffices to destroy any coherence that would define spreading wave packets for their centers of mass (see Fig. 4.2). If the wavelengths of the abundant scatterers are larger than the width of the wave packet, an otherwise free 'particle' is dynamically described by the master equation (Joos and Zeh 1985)

$$i\frac{\partial\rho(x,x',t)}{\partial t} = \frac{1}{2m}\left(\frac{\partial^2}{\partial x'^2} - \frac{\partial^2}{\partial x^2}\right)\rho - i\lambda(x-x')^2\rho. \qquad (4.35)$$

It can be *derived* from a universal Schrödinger equation by assuming the dynamical irrelevance of all correlations with the environment after they have formed, and by neglecting recoil (see also Chap. 3 and Appendix 1 of Joos et al. 2003). The coefficient λ is here determined by the rate of scattering and its efficiency in orthogonalizing states of the environment. In the small wavelength limit, a *single* collision is usually sufficient to destroy any coherence beyond the wavelength. The decoherence rate is then simply given by the scattering rate (that is, the product of the flux of environmental particles and the total cross-section). Even the interpretation of the wave mechanical scattering process as consisting of individual collision *events* can be explained by further decoherence of superpositions of different 'collision times' in a process that is actually smooth (see Sect. 4.3.6).

So one does not have to *postulate* a fundamental semigroup in order to describe open quantum systems (Sect. 4.4). If the environment forms a heat bath, (4.35) describes the infinite-mass limit of *quantum Brownian motion*

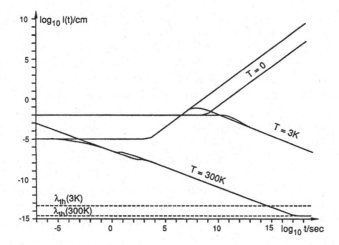

Fig. 4.2. Time dependence of the coherence length $l(t)$ for the center of mass of a small dust grain of 10^{-14} g with radius 10^{-5} cm under continuous measurement by thermal radiation. The six curves represent two initially pure Gaussian wave packets, differing by their initial widths $l(0)$, and three different temperatures T of the radiation. $T = 0$ describes the free dispersion of the wave packet according to the Schrödinger equation, which holds otherwise as an approximation for a limited time only. Scattering of atoms and molecules is in general far more efficient than that of thermal photons – even in intergalactic space. Brownian motion becomes relevant only when the coherence length approaches the de Broglie wavelength λ_{th}. From Joos and Zeh (1985)

(see Caldeira and Leggett 1983, Zurek 1991, Hu, Paz and Zhang 1992, Omnès 1997). This demonstrates that, even for entirely negligible recoil (which would be responsible for noise and friction), there remains an important effect that is based on quantum nonlocality.

Apparent classical properties thus *emerge* from the wave function, and are maintained, by a process that cannot be reversed. In particular, particle aspects (such as tracks in a bubble chamber) arise in the form of macroscopic phenomena (bubbles) which are observed at certain positions in space because of their decoherence. Similarly, the disappearance of interference between *partial waves* in a *Welcher Weg* measurement (Scully, Englert and Walther 1991) does not require any wave–particle 'complementarity'. Furthermore, no super-luminal tunnelling (see Chiao 1998) may occur according to a consistent quantum description, since *all parts* of a wave packet propagate (sub-)luminally, while its group velocity does not represent the propagation of any physical objects.

Master equations for open systems can also be derived by means of the *decoherence functional* (Feynman and Vernon 1963, Mensky 1979). Feynman's path integral is thereby used as a *tool* for calculating the propagation of a global density matrix, while the environment is again continuously traced out when getting entangled with the considered system. The intuitive picture of an *ensemble of paths* (representing different *possible trajectories*) is justified only if this superposition of paths decoheres into narrow wave packets. A 'restriction' of the path integral by the presence of absorbers (Mensky 2000) would be equivalent to a corresponding reduction of the (total) wave function.

All quasi-classical phenomena, including those representing apparently reversible (friction-free) mechanics, rely conceptually on irreversible decoherence. This requires the continuous production of objective physical entropy (increasing entanglement), which may be macroscopically negligible, but is large in terms of bits. If the quasi-classical trajectories are chaotic, this entropy production may be controlled by the classical Lyapunov exponent (Zurek and Paz 1994, Monteoliva and Paz 2000), even though the entanglement entropy does not require any (initial) uncertainties that would grow *in the direction of calculation*, as assumed in the classical theory of chaos (see Sect. 3.1.2).

General Literature: Joos's Sect. 3.2 of Joos et al. (2003), Hornberger, Hackermüller, and Arndt (2005).

4.3.2 Molecular Configurations as Robust States

Chirality of molecules, such as right- or left-handed sugar, represents a discrete elementary variable controlled by decoherence. Although a chiral state is described by a certain wave function, it is not an energy eigenstate, which would have to be a parity eigenstate, that is, a symmetric or antisymmetric superposition of both chiralities (see Zeh 1970, Primas 1983, Woolley 1986). The reason is that it is chirality (not parity) that is continuously 'measured', for example by scattered air molecules – in analogy to position rather than momentum being measured for a macroscopic 'mass point'. For sugar molecules under normal conditions, the decoherence time scale is of the order of 10^{-9} s (Joos and Zeh 1985), while the tunneling time between different chirality states is extremely long.

As a consequence, each individual molecule in a bag of sugar retains its chirality, while a parity state – if it had come into existence in a mysterious or expensive way – would almost immediately 'collapse' into an apparent ensemble of two chirality states (with equal probabilities). Parity would thus *not be conserved* for sugar molecules, while chirality is always confirmed when measured twice (although it is not a constant of the motion).

This *robustness* against decoherence seems to characterize properties that we usually regard as 'elements of classical reality', such as spots on the photographic plate or other 'pointer states' of a measurement device. Discrete states may even be protected against otherwise possible transitions (tunneling) by the *quantum Zeno effect*. For continuous variables, the concept of robustness is compatible with a (regular) time dependence according to a master equation, as described in the previous section for the quasi-classical center of mass motion of macroscopic objects. Since entropy production by interaction with the environment is lowest for a density matrix that is already diagonal in terms of robust states, this property has been called a 'predictability sieve', and proposed as a definition of classical states (Zurek, Habib and Paz 1993).

Dynamical robustness is also essential for the physical concept of memory or information storage, such as in DNA, brains or computers. Even 'states-of-being-conscious' (see Chap. 1) seem to be quasi-classical in this sense (Tegmark 2000) – at least inasmuch as they are able to communicate. In contrast to such robust properties, which can be assumed to exist regardless of their actual measurement, *potentially* measurable quantities have been called 'counterfactuals'. Their superpositions, which would themselves describe individual physical states, must not be assumed to describe ensembles of definite (really existing though unknown) properties. Such different concepts of reality (operational or phenomenological versus hypothetical though consistent and economically chosen) can thus be analyzed and understood in terms of decoherence, which is thereby assumed to represent a physical process in a consistent (nonlocal) quantum reality, while elements of phenomenological (classical) reality 'emerge' (or become 'factual') only under certain environmental conditions. If these conditions may change, such as for microscopic systems under different measurements, the emerging concepts naturally vary between 'complementary' modes of description.

Chemists know furthermore that atomic nuclei or strongly bound ions as constituents of *large* molecules have to be described classically (for example as quasi-rigid configurations, which may vibrate or rotate in a time-dependent manner), while the electrons have to be described by stationary or adiabatically comoving wave functions. This asymmetric behavior is often attributed, by means of a Born–Oppenheimer approximation, to their large mass ratio. However, this argument is insufficient, since this approximation applies as well to small molecules that are found in discrete energy eigenstates, which are completely described by stationary wave functions, giving rise to discrete rotational and vibrational energy bands rather than quasi-classical states.

The formation of time-dependent (particle-like) wave packets for the atomic nuclei in large molecules can instead be understood once again by means of decoherence (Joos and Zeh 1985). For example, the positions of nuclei are usually permanently monitored by scattering of lighter molecules that form the environment. But why only the nuclei (or ions), and why not very small molecules? The answer requires a quantitative investigation in each individual case, and the result depends on a delicate balance between internal dynamics and interaction with the environment, whereby the density of states plays a crucial role (Joos 1984). This may then lead approximately to either (a) unitary evolution (including stationary states), (b) a master equation, or (c) freezing of the motion (quantum Zeno effect). Much numerical work remains to be done for such complex systems, while simple ones may be described by an effective master equation, such as (4.35), for example.

General Literature: Joos's Sect. 3.2.4 of Joos et al. 2003.

4.3.3 Quantum Computers

Digital computers are based on robust binary states, carrying 'bits' of information. Even neural networks can be described to some extent by states of cells having 'fired' or not, while DNA is based on four different 'letters', each one therefore representing two bits. Chiral molecules also represent bits, although they would not be very convenient for information handling.

Just as chiral states may be robust because of their decoherence, so are all macroscopic constituents that are used in classical computers. However, on a microscopic scale there also exist quantum bits (or 'qubits'), which may occur in all conceivable superpositions of their two basic states. In some cases, such as photon polarizations or spinors that may form spin lattices, they may even be assumed to be isolated from the environment to a good approximation. Such isolated qubits form the essential constituents of quantum computers. Because of their greater variety of possible states (for example spin-up and spin-down in any direction of space), and the possibility of getting entangled, they offer quite novel possibilities for computing (see Shor 1994).

The problem here is that completely isolated systems, required for a unitary evolution, could hardly be manipulated or read as wished for a usable computer. On the other hand, any uncontrollable effect of the collective state of an n-qubit system on the environment would immediately destroy (that is, irreversibly dislocalize) the crucial superposition that forms the state of this system as a whole. This vulnerability of quantum computers against decoherence grows exponentially with their size, so that macroscopic quantum computers may have to be excluded by superselection rules, similarly to macroscopic superpositions in general. Superpositions containing a large number of entangled electrons that have been prepared and observed in the laboratory (Mooij et al. 1999, Friedman et al. 2000) are facilitated by 'freezing out' most of the degrees of freedom in a degenerate state – in stark contrast to what would be required for the complexity of a a quantum computer.

In an attempt to overcome this problem, various *correction codes* have been proposed (see Bouwmeester, Ekert, and Zeilinger 2000). They are conventionally based on some concept of multiple redundancy (an internal kind of back-up), that would have to further enlarge the number of qubits. However, while redundancy may be used as a protection against distortions of the computer by the environment, decoherence is a distortion of the environment by the computer. It can only be corrected for inasmuch as the environment remains controllable – certainly not a very realistic assumption. Usable quantum computers may therefore be excluded in practice for some time to come (see also Haroche and Raimond 1996). It would be quite inconsistent, though, to study the possibility of quantum computers even in principle, while at the same time denying the *reality* of all components of a quantum superposition or wave function – as appropriately emphasized by David Deutsch (1997). Decoherence, too, is the consequence of such an assumption.

In order to give rise to a *classical* computer, each bit would have to be decohered after each calculational step. This would produce precisely the minimum amount of entropy of $k \ln 2$ that was conjectured to be required by Landauer for other reasons (see the end of Sect. 3.3), but then refuted by Bennett in a classical deterministic setting. This entropy production would thus again (have to) be avoided in quantum computers according to the deterministic Schrödinger equation, which is valid only for isolated systems.

General Literature: Bouwmeester, Ekert, and Zeilinger (2000).

4.3.4 Charge Superselection

Gauss' law, $q = (1/4\pi) \int \boldsymbol{E} \cdot \mathrm{d}\boldsymbol{S}$, tells us that every local electric charge requires a certain flux of electric field lines through a sphere surrounding it at any distance. For a superposition of different charges, one would therefore obtain an entangled quantum state of charges and fields,

$$\sum_q c_q \psi_q^{\text{total}} = \sum_q c_q \chi_q \Psi_q^{\text{field}} = \sum_q c_q \chi_q \Psi_q^{\text{near}} \Psi_q^{\text{far}}$$

$$=: \sum_q c_q \chi_q^{\text{dressed}} \Psi_q^{\text{far}} , \qquad (4.36)$$

where χ_q represents the bare charge, while $\Psi_q^{\text{field}} = \Psi_q^{\text{near}} \Psi_q^{\text{far}}$ is the state vector of its correlated electrostatic field, symbolically written as a tensor product of a near field and a far field (see Sect. 2.3). The dressed (physical) charged particle would then be described by a density operator of the form

$$\rho_{\text{local}} = \sum_q |\chi_q^{\text{dressed}}\rangle |c_q|^2 \langle \chi_q^{\text{dressed}}| , \qquad (4.37)$$

provided that the states of the far field for different charge q are mutually orthogonal (uniquely distinguishable). The charge is thus decohered by its own Coulomb field, and no charge superselection rule has to be *postulated* (see Giulini, Kiefer and Zeh 1995). The formal decoherence of the bare charge by its near field remains unobservable, since experiments can only be performed with dressed charges.

While this result explains the observed charge superselection rule, one may ask what it means locally. What if an electric charge is accompanied by a negative one at a different place? Or at what distance and on what time scale would the superposition of two different locations of a point charge (such as those of an electron during an interference experiment) be decohered by the quantum state of the corresponding dipole field. A *classical* retarded Coulomb field would contain causal information about the precise path of its source particle. However, interference between different paths of an electron has been demonstrated to exist at least over distances of the order of millimeters (Nicklaus and Hasselbach 1993). This indicates that the Coulomb field contributes to decoherence only by its monopole component, sufficient to explain charge superselection.

This conclusion can indeed be understood in terms of quantum theory, since photons with diverging wavelength (which may be regarded as representing static fields) cannot distinguish different charge positions – even though the number of such virtual photons would diverge in a Coulomb field. Static dipole (or higher) multipole moments do not possess any far fields. Therefore, only the 'topological' Gauss constraint $\partial_\mu F^{\mu 0} = 4\pi j^0$ contributes to the decoherence of the physical particle by the Coulomb field. Any time-dependence (including a retardation) must then be described in terms of transverse photons, represented by the vector potential $\boldsymbol{A}$ (with div $\boldsymbol{A} = 0$ in the Coulomb gauge). In this picture, only the spatial distribution of electric field lines – not their total flux – forms dynamical degrees of freedom that have to be quantized. Charge decoherence has therefore been regarded as 'kinematical', although it might as well be assumed to be dynamically *caused* by the retarded field of the (conserved) charge in its past – or equivalently by the advanced field resulting from its future. Note, however, that a kinematical Coulomb constraint is in conflict with the concept of a physical Hilbert space that is spanned by direct products of *local* states.

Dipoles and higher moments (which can define position *differences* for a point charge), can thus be measured by the environment either through emission (or scattering) of transverse ('real') photons, or by the irreversible polarization of nearby matter (Kübler and Zeh 1973, Anglin and Zurek 1996). The latter effect has now been experimentally confirmed (Sonnentag and Hasselbach 2005). In general, this decoherence is not 'pure', but related to energy transfer, although the recoil caused by emission of soft photons may be negligible. The (often virtual) decoherence of individual charged particles *within* solid bodies is discussed in Imry (1997).

The emission of photons would require the charge to be *accelerated*. For example, a *transient dipole* of charge e and maximum distance d, caused by spatially separating opposite charges for a time interval t, requires accelerations a of the order d/t^2. According to Larmor's classical formula (see Sect. 2.3), the intensity of radiation is then at least $2e^2a^2/3$. In order to resolve the position difference, the emitted radiation has to consist of photons with energy greater than $\hbar c/d$ (that is, wavelengths smaller than d). The probability that information about the dipole is radiated away by at least one photon is then very small: of order $\alpha Z^2 (d/ct)^3$, where α is the fine structure constant and Z the charge number. In more realistic cases, such as interference experiments with electrons, stronger accelerations may occur, but they would in general still cause negligible decoherence.[4] Decoherence of the position of a charged

---

[4] This limitation of the information capacity of an electromagnetic field by its quantum nature must also give rise to an upper bound for the validity of Borel's argument of Sect. 3.1.2.

particle is therefore dominated by *scattering* of photons, and by interaction with charged or polarizable matter.

The *gravitational field* of a point mass is similar to the Coulomb field of a point charge. Superpositions of different mass should therefore be decohered by the quantum state of the monopole contribution of spatial curvature, and thus give rise to a mass superselection rule. However, superpositions of different energies (hence masses) evidently exist, since they form the time-dependent states of local systems. This situation may not yet be sufficiently understood.

The Coulomb field would vanish globally if the total charge of the Universe were zero (see Giulini, Kiefer and Zeh 1995). This would eliminate the need for a Gauss constraint for the Universe. The gravitational counterpart of this global consequence is the absence of time from a closed Universe in quantized general relativity (the Hamiltonian constraint – see Sect. 6.2).

General Literature: Kiefer's Sect. 4.1.1 and Giulini's Chap. 6 of Joos et al. 2003.

4.3.5 Quasi-Classical Fields and Gravity

Not only are the quantum states of charged particles decohered by their fields – quantum states of fields may in turn be decohered by the currents on which they act. In this case, 'coherent states', that is, Schrödinger's time-dependent but dispersion-free Gaussian wave packets for the amplitudes of classical wave modes (eigenmodes of coupled oscillators), have been shown to be robust for similar reasons as electric charges, chiral molecules or the wave packets describing the center of mass motion of quasi-classical objects (Kübler and Zeh 1973, Kiefer 1992, Zurek, Habib and Paz 1993, Habib et al. 1996). This explains why macroscopic states of neutral boson fields appear as *classical fields*, and why superpositions of macroscopically different 'mean fields' or different vacua (Sect. 6.1) are never observed.

Coherent harmonic oscillator states, which form states of minimum Heisenberg uncertainty, can be defined (for each wave mode k) as eigenstates $|\alpha_k\rangle$ of the non-Hermitean annihilation (or energy-lowering) operators a_k with their complex eigenvalues α_k (that is, $a_k|\alpha_k\rangle = \alpha_k|\alpha_k\rangle$). These Gaussian wave packets are centered at a time-dependent classical field amplitude $\alpha_k(t) = \alpha_k^0 e^{i\omega t}$, where $\text{Re}(\alpha_k)$ and $\text{Im}(\alpha_k)$ represent the electric and magnetic field strengths, formally equivalent to the position and momentum of a mechanical oscillator. Since the interaction between the field and its charged sources is usually linear in the field operators a_k or $a_k^\dagger$, these coherent states form an (overcomplete) robust 'pointer basis': they create minimal entanglement with their 'environment' (that consists here of charged sources that happen to be present).

In contrast to these superpositions of many different photon numbers (or oscillator quantum numbers), *single*-photon states resulting from the decay of different individual atoms (or even the n-photon states resulting from the decay of a *different number* n of atoms) are unable to interfere with one another, since they are entangled with mutually orthogonal final states of the sources. Two incoherent components of a one-photon state may then appear as 'different' photons (using Dirac's language), although the photons themselves are indistinguishable. A quasi-classical collective state of the source, however, would hardly change (judged in terms of the Hilbert space inner product) when emitting a photon. It is thus able to *produce* the coherent superpositions of different photon numbers discussed above (see also Kiefer 1998).

Although the coherent states behave macroscopically, superpositions of *different* ones, $c_1|\alpha_1\rangle + c_2|\alpha_2\rangle$ (called 'Schrödinger cat states'), have been produced and maintained for a short time as one-mode laser fields in a cavity (Monroe et al. 1996). These mesoscopic superpositions must decohere, similarly to a Schrödinger cat, although on a time scale that is slow enough to allow this decay of coherence to be monitored as a function of time. In this way, decoherence was for the first time confirmed experimentally as a smooth process in accord with the Schrödinger equation (Davidovich et al. 1996, Brune et al. 1996).

Arguments similar to those used in quantum electrodynamics (QED) apply to *quantum gravity* (Joos 1986, Kiefer 1999 – for applications to quantum cosmology see Chap. 6). Quantum states of matter and geometry must be entangled, and give rise to mutual decoherence. The classical appearance of spacetime geometry is thus no reason *not* to quantize gravity. The beauty of Einstein's theory can hardly be ranked so much higher than that of Maxwell's to justify its exemption from quantization. An exactly classical gravitational field interacting with a quantum particle would be incompatible with the uncertainty relations – as has been known since the early Bohr–Einstein debate. The reduced density matrix for the metric must therefore be expected to represent an apparent mixture of different quasi-classical curvature states. Since the observer cannot avoid being correlated to them, spacetime curvature always appears to be classically given – see Sects. 4.6 and 6.2.

Moreover, the entropy and thermal radiation (of all fields) characterizing a black hole or an accelerated Unruh detector (Sects. 5.1 and 5.2) are consequences of the entanglement between relativistic vacua on two half-spaces separated by a horizon (each one forming the environment of the other). This entanglement entropy measures the same type of 'apparent' ensemble as the entropy produced according to the master equation (4.35) for a macroscopic mass point. The disappearance of coherence behind a horizon has nonetheless occasionally been regarded as a *fundamental* violation of unitarity, and even as the ultimate source of irreversibility (see Sects. 4.4, 5.1 and 6.2). This appears neither justified nor required (see Kiefer, Müller and Singh 1994, Kiefer 2007, Zeh 2005a).

General Literature: Kiefer's Chap. 4 of Joos et al. 2003, Kiefer 2004.

4.3.6 Quantum Jumps

Quantum objects are often observed by means of flashes on a scintillation screen or 'clicks' of a counter. These macroscopic phenomena are then interpreted as caused by pointlike objects, passing through the observing instrument during a short time interval, while this is in turn understood as evidence for a discontinuous 'decay event' (for example, of an atomic nucleus). A *rate equation* for such events is equivalent to a master equation, while a *constant* relative rate would describe exponential decay of the source. Discrete quantum jumps between two energy eigenstates have even been observed for single atoms in a cavity by permanently monitoring their energy, thus enforcing decoherence between energy eigenstates (Nagourney, Sandberg and Dehmelt 1986, Sauter et al. 1986, Pegg, Loudon and Knight 1986, Gleyzes et al. 2006). Therefore, formal creation and annihilation operators are often misunderstood as defining discrete events, even though they occur in a Hamiltonian that constitutes a Schrödinger equation.

This Schrödinger equation would describe a state vector that smoothly develops components with different particle numbers, or a wave function that leaks out of an unstable system (such as a quantum 'particle' in a potential well). This contrast between discrete events and the Schrödinger equation is clearly the empirical root of the probability interpretation of the wave function in terms of events and particles. A wave function can exponentially decay only in a limited region of space (for example within an expanding sphere for a limited time – see Sect. 4.5). This wave function is a *superposition* rather than an ensemble of different decay times. Their interference and the dispersion of the corresponding outgoing wave lead to deviations from an exponential decay law. Although these deviations are too small to be observed for decay into infinite space, interference between different decay times has often been confirmed in other situations, not least as 'coherent state vector revival' for photons emitted into cavities with reflecting walls (Rempe, Walther and Klein 1987).

In Sect. 4.3.1, the appearance of particles following tracks in a cloud chamber has been explained in terms of an apparent ensemble of narrow wave packets arising by means of decoherence. Similar arguments may as well explain apparently discrete events. Even if quantum objects remain isolated before being detected, they would be decohered in the detector – usually on a very short time scale. Therefore, the same decoherence that describes localization in space also explains localization in time. Jumps between discrete energy levels, observed under continuous measurement, represent apparently discrete 'decay histories', which can be explained by Mott-type quantum correlations between successive measurements of short but finite individual duration (including the decoherence of their outcomes). Neither particles nor genuine quantum jumps are required as *fundamental* concepts in quantum theory (Zeh 1993, Paz and Zurek 1999). Whenever decay fragments (or the decaying object) interact appropriately with their environment, interference between two partial waves describing a decayed state and a not yet decayed state disappears on a very short (though finite) decoherence time scale, thus giving rise to an apparent *ensemble* of decay times. This time scale is in general much shorter than the time resolution of measurements.

If the decay status is thus permanently 'monitored' by the environment, a set of identical decaying objects is thus more appropriately described by a rate equation than by a Schrödinger equation (Sect. 4.5. This rate equation leads to an exact exponential law, since it excludes any interference between different decay times. Similarly, decay products emitted in superpositions of sufficiently different energies are absorbed into mutually orthogonal final states of the environment. Microscopic systems with their discrete energy levels must therefore decohere into eigenstates of their own Hamiltonians. This explains why the atomic world is characterized by stationary states, and von Neumann spoke of an *Eingriff* (intervention) required for their change.

So it seems that this situation of continuously monitored decay has led to the myth of quantum theory as a stochastic theory for fundamental *quantum events* (see Jadczyk 1995). Bohr (1928) remarked that "the essence" (of quantum theory) "may be expressed in the so-called quantum postulate, which attributes *to any atomic process* an essential discontinuity, or rather individuality ..." (my italics). This statement is in conflict with many microscopic and mesoscopic quantum phenomena that have since then been observed. Heisenberg and Pauli similarly emphasized their preference for matrix mechanics because of its (evidently misleading) superiority in describing discontinuities. Ole Ulfbeck and Aage Bohr (2001) recently emphasized the unpredictable occurrence of 'clicks in the counter', while denying the existence of any quantum events in the source that would precede them. This comes close to the consequences of decoherence, but rather than taking into account entanglement with the environment the authors conclude that "the wave function then loses its meaning". According to the decoherence theory, the underlying entanglement processes are always smooth, and described by a Schrödinger equation. The short decoherence time scales lead to the impression of quantum *jumps* between energy eigenstates, for example, while narrow wave packets are interpreted as particles or classical variables (even though the certainty of classical properties has to be restricted by the uncertainty relations in order to comply with the Fourier theorem).

While the description of all physical phenomena in terms of time-dependent entangled wave functions now appears as a consistent picture, an important question remains: how should the probabilities, which were required to justify the concept of a density matrix in Sect. 4.2, be understood if they are *not* probabilities for quantum jumps or for the occurrence of measurement results in the form of fundamental 'events'. This discussion will be resumed in Sect. 4.6.

General Literature: Joos's Sect. 3.4.1 of Joos et al. 2003.

4.4 Quantum Dynamical Maps

Open systems can be phenomenologically described by means of semigroups – thus *postulating* an arrow of time. In quantum theory, they possess some novel aspects in comparison to their classical counterparts (Sect. 3.4). For example, these 'quantum dynamical maps' have been used to formalize von Neumann's 'first intervention' (the reduction of the wave function) as part of the dynamics (Kraus 1971). This is possible, since semigroups can not only describe the transition of pure states into ensembles, but also the 'selection' of an *individual* element from them (see below). Otherwise they are equivalent to an entropy-enlarging Zwanzig-type master equation with respect to the corresponding $\hat{P}_{\mathrm{sub}}$. Although 'irrelevant' correlations with the environment, which would arise according to the unitary global dynamics, now represent quantum entanglement, they are usually not distinguished from classical statistical correlations when it comes to applications.

This confusion of concepts is equivalent to a popular but insufficient 'naive' interpretation of decoherence, which pretends to derive genuine ensembles. Quantum dynamics is occasionally even *defined* in terms of semigroups, assumed to act on the density matrix as a fundamental kinematical object characterizing quantum systems. (Hence the term 'statistical operator' for the density operator.) However, this 'minimal statistical interpretation' entirely neglects the difference between genuine and apparent ensembles, and thus all consequences of entanglement beyond the considered systems (quantum nonlocality). Even the superposition principle has been claimed to be derivable in this formalism (Ludwig 1990), although it is then simply reintroduced in a hidden form (for example by changing the laws of statistics in an unjustified way).

Semigroups are certainly mathematically elegant and powerful. Therefore, they would form candidates for *new* theories if conventional (Hamiltonian) quantum theory should prove wrong empirically as a universal theory. The question is whether mathematical elegance already warrants physical relevance or is merely convenient within a certain approximation. To quote Lindblad (1976): "It is difficult, however, to give physically plausible conditions ... which rigorously imply a semigroup law of motion for the subsystem. ... Applications ... have led some authors to introduce the semigroup law as the fundamental dynamical postulate for open (non-Hamiltonian) systems." Such a law would *fundamentally* introduce an arrow of time (see also Sect. 4.6), but it would depend on the choice of 'systems' – and may be at variance with certain experiments which confirm quantum nonlocality.

The simplest quantum systems (such as spinors) are described by a two-dimensional Hilbert space. Their density matrix may be written by means of the Pauli matrices σ_i ($i = 1, 2, 3$) in the form

$$\rho = \frac{1}{2}(1 + \boldsymbol{\sigma} \cdot \boldsymbol{\pi}) \,, \tag{4.38}$$

where the (mathematically) real *polarization vector* $\boldsymbol{\pi} = \mathrm{Trace}\{\boldsymbol{\sigma}\rho\}$ – that is, the expectation value of *all* spin components – completely defines ρ as a general Hermitean 2×2 matrix of trace 1. The latter is in turn equivalent to a (genuine or apparent) *ensemble* of spinors. The length of $\boldsymbol{\pi}$ is a measure of the purity of the 'mixed state' ρ, since $\mathrm{Trace}\{\rho^2\} = (1 + \boldsymbol{\pi}^2)/2$, with $\boldsymbol{\pi}^2 \leq 1$. A pure state corresponds to a unit polarization vector, while an arbitrary density matrix (a general 'state' in the language of mathematical physics) is characterized by the mean value $\boldsymbol{\pi} = \sum_\alpha p_\alpha \boldsymbol{\pi}_\alpha$ of all unit vectors $\boldsymbol{\pi}_\alpha$ in an ensemble of spinors that may represent this density matrix.

A general trace-preserving linear superoperator $\hat{P}$ acting on ρ must be defined on 1 and $\boldsymbol{\sigma}$ in order to be completely defined:

$$\hat{P}1 := 1 + \boldsymbol{\pi}_0 \cdot \boldsymbol{\sigma} , \qquad \hat{P}\boldsymbol{\sigma} := \boldsymbol{A} \cdot \boldsymbol{\sigma} , \tag{4.39}$$

with a real vector $\boldsymbol{\pi}_0$ and a linear vector transformation $\boldsymbol{A}$. $\hat{P}$ is idempotent (a Zwanzig 'projector') if $\boldsymbol{A}^2 = \boldsymbol{A}$ and $\boldsymbol{\pi}_0 \cdot \boldsymbol{A} = 0$ ($\boldsymbol{A} = 0$, for example). If $\boldsymbol{\pi}_0 \neq 0$, $\hat{P}$ creates new information – even from the unit matrix (see Sect. 3.2).

Dynamical combination of the projection $\hat{P}$ with a Hamiltonian evolution (which would describe a rotation of $\boldsymbol{\pi}$) in the form of a master equation leads to the *Bloch equation* for the vector $\boldsymbol{\pi}(t)$,

$$\frac{\mathrm{d}\boldsymbol{\pi}}{\mathrm{d}t} = \boldsymbol{\omega} \times (\boldsymbol{\pi} - \boldsymbol{\pi}_0) - \sum_i \gamma_i (\pi^i - \pi_0^i) \boldsymbol{e}_i , \tag{4.40}$$

in a certain vector basis $\{\boldsymbol{e}_i\}$ (see Gorini, Kossakowski and Sudarshan 1976). Values of $\gamma_i < 0$ or $|\boldsymbol{\pi}_0| > 1$ would violate the positivity of the density matrix at some $t > 0$, and thus have to be excluded.[5] The second term on the RHS describes anisotropic 'damping' towards $\boldsymbol{\pi}_0$. This formal creation of information (which is contained in the vector $\boldsymbol{\pi}_0$) may represent very different physical situations, such as equilibration with an external heat bath of given (possibly lower) temperature, or evolution towards a certain measurement result. The density matrix defined by the polarization vector $\boldsymbol{\pi}_0$ is often called a *reference state*, while the *relative entropy* with respect to it [see (3.55)] never decreases under the Bloch equation – even when the physical entropy of the local spinor system does. Hermiticity of $\hat{P}$ (corresponding to a genuine, entropy-raising projection operator) would require $\boldsymbol{\pi}_0 = 0$ and $\boldsymbol{A} = \boldsymbol{A}^\dagger$, that is, a projection of the vector $\boldsymbol{\pi}$ onto a specific component.

If the two-dimensional Hilbert space describes something other than spin, such as isotopic spin, a $K, \bar{K}$ system, or fermion occupation numbers, the

---

[5] As mentioned in Sect. 4.2, *all* subsystem density matrices remain positive under global Hamiltonian dynamics, and even under a collapse of the global state vector. This property of 'complete positivity' has to be separately *postulated* for phenomenological quantum dynamical maps (see Kraus 1971), thus further illustrating the fact that these maps *cannot* be regarded as representing fundamental physics.

polarization vector lives in an *abstract* three-dimensional space that usually cannot simply be 'rotated' in practice. The abstract formalism can also be generalized to n-dimensional Hilbert spaces. For this purpose the Pauli matrices have to be replaced by the $(n^2 - 1)$ Hermitean generators of the corresponding group $SU(n)$, while the real 'coherence vectors' (the generalizations of the polarization vector $\boldsymbol{\pi}$) now live in the vector space spanned by them. For example, SU(3) gives rise to the 'eight-fold way'. The most important *new* property then is that there are more than one (in fact, $n-1$) *commuting* Hermitean generators. They may contain a nontrivial subset that is decohered under all realistic environmental conditions, and thus may form the center of a phenomenological set of observables (the set of 'classical observables' – see Sect. 4.3). For example, maps of density matrices of dimension $n = 4$ which happen to completely decohere with respect to a certain basis may reproduce the classical maps of Figs. 3.8a, c and d.

In the infinite-dimensional Hilbert space of quantum mechanics, the *Wigner function*

$$
\begin{aligned}
W(p, q) &:= \frac{1}{\pi} \int e^{2ipx} \rho(q + x, q - x) \mathrm{d}x \\
&\equiv \frac{1}{2\pi} \iint \delta\left(q - \frac{z + z'}{2}\right) e^{ip(z - z')} \rho(z, z') \mathrm{d}z \, \mathrm{d}z' \\
&= \mathrm{Trace}\{\Sigma_{p,q} \rho\} \,,
\end{aligned} \tag{4.41}
$$

where the third line is written in analogy to $\boldsymbol{\pi} = \mathrm{Trace}\{\boldsymbol{\sigma}\rho\}$, assumes the role of the coherence vector $\boldsymbol{\pi}$. Evidently,

$$
\Sigma_{p,q}(z, z') := \frac{1}{2\pi} e^{ip(z - z')} \delta\left(q - \frac{z + z'}{2}\right) \tag{4.42}
$$

is a generalization of the Pauli matrices (with the vector index replaced by p, q). On a finite q-interval of length L, $\Sigma_{p,q}$ would require an additional term $-(1/2\pi L)e^{ip(z - z')}$ in order to warrant tracelessness.

The Wigner function is thus a continuous set of expectation values of these generalized Pauli matrices. They form the components (one for each phase space point) of a generalized coherence vector. This 'vector' characterizes the density matrix ρ completely – just as in (4.38) and regardless of its interpretation. Although it does *not* represent a probability density on phase space, as illustrated by its possibly negative values, one may calculate all expectation values in the *form* of a mean value for a classical function of state $f(p, q)$, viz., $\langle F \rangle = \int f(p, q) W(p, q) \mathrm{d}p \, \mathrm{d}q$. However, a 'quasi-classical' Gaußian wave packet, for example, is a coherent quantum mechanical superposition of position or momentum eigenstates in spite of its (in this case) non-negative Wigner function.

Lindblad (1976) was indeed able to generalize the Bloch equation to infinite-dimensional Hilbert spaces. He wrote it (in a form that applies to the density matrix) as

$$\mathrm{i}\frac{\partial\rho}{\partial t} = [H, \rho] - \frac{\mathrm{i}}{2}\sum_k \left(L_k^\dagger L_k \rho + \rho L_k^\dagger L_k - 2L_k \rho L_k^\dagger\right) , \qquad (4.43)$$

with arbitrary generators L_k in Hilbert space. It describes a *creation* of information, that is, a local decrease of the corresponding von Neumann entropy, if and only if some generators do not commute with their Hermitean conjugates $L_k^\dagger$. This can be shown by applying the non-Hamiltonian terms of (4.43) to the unit matrix $\rho = 1$, which describes *no* information. Otherwise this equation describes information loss (a genuine Zwanzig projection).

This can also be seen by means of the general representation of a Zwanzig projector on density operators in quantum mechanical Hilbert space, $\hat{P}\rho = \sum_k V_k \rho V_k^\dagger$. It is similar to the square root of a positive operator, written in its eigenbasis. If $L_k^\dagger = L_k$, the non-Hamiltonian Lindblad terms assume the form of a double commutator, $L^2\rho + \rho L^2 - 2L\rho L = \left[L, [L, \rho]\right]$. For $L = \sqrt{2\lambda}x$ one recovers (4.35), that is, decoherence in the x-basis, precisely as *derived* from unitary interaction with the environment.

In this way one may, in particular, describe transitions of pure states into formal ensembles of measurement results with their corresponding Born probabilities. However, 'damping' towards a definite *pure* state (a semigroup proper in the sense of Fig. 3.8c), would require the second term of (4.40) with a unit vector π_0. It allows one to describe the evolution into a (freely chosen) definite measurement outcome. This dynamics can then readily be combined with a stochastic formalism that is defined to select the possible final states of a measurement in accordance with the Born rules (Bohm and Bub 1966, Pearle 1976, Gisin 1984, Belavkin 1988, Diósi 1988). If applied continuously, such as by means of the Itô process, this formalism describes a genuine collapse as a smooth but indeterministic process (Pearle 1989, Ghirardi, Pearle and Rimini 1990). If this modification of the Schrödinger equation were correct, it should in principle be observable, although it would usually be camouflaged by environmental decoherence (Joos 1986, Tegmark 1993).

Many explicit collapse models of this kind have been proposed in the literature. Some remain ambiguous about their true intentions (that is, whether they are meant fundamental or phenomenological), or simply disregard the difference between genuine and apparent ensembles (proper and improper mixtures). In particular, the *quantum state diffusion model* (Gisin and Percival 1992) presumes that reduced density matrices can be 'untangled' into genuine ensembles (with only one of their members assumed to represent reality). However, this would again be equivalent to a modification of the global Schrödinger equation (Diósi and Kiefer 2000).

General Literature: Alicki and Lendi 1987, Diósi and Lukács 1994, Stamatescu's Chap. 8 of Joos et al. 2003.

4.5 Exponential Decay and 'Causality' in Scattering

There are only a few absolutely stable 'particles' (elementary quantum objects), while all others are known as decaying on vastly different time scales. In quantum theory, they may be described formally by means of *complex energies*. For a Schrödinger type time dependence e^{-iEt}, a negative imaginary part, $E = E_0 - i\gamma$ with $\gamma > 0$, would lead to an exponentially decreasing wave function. This does not just describe probabilities for different decay times, since all parts of the wave function form *one* coherent superposition (see below and Sect. 4.3.6). Even though microscopic, these objects have to be regarded as *open* quantum systems. For example, an excited atom is coupled to an initial vacuum (or a photon heat bath of zero temperature). Unbounded space represents an 'absorber' of infinite capacity for the decay fragments.

The decaying system may also be described by means of an S-matrix for the decay fragments, where unstable states show up as poles in the complex energy or momentum plane. This S-matrix must represent the fundamental (time-symmetric) dynamics. Exponential decay then seems to characterize a fundamental direction in time (see Prigogine 1980, for example), similar to Ritz's retarded electrodynamics (Chap. 2). Since there are no energy eigenstates with complex eigenvalues (sometimes called 'Gamow vectors') in Hilbert space, this situation has even led to the proposal of 'rigged Hilbert spaces' (Böhm 1978). However, decaying systems may well be described in conventional quantum mechanical terms, where the exponential time dependence applies only approximately in a limited spacetime region.

Exponential decay of an arbitrary quantity A would be the consequence of a constant loss rate, described by

$$\frac{\mathrm{d}A}{\mathrm{d}t} = -\lambda A \,, \tag{4.44}$$

with $\lambda > 0$. The absolute rate of change, $\mathrm{d}A/\mathrm{d}t$, is then completely determined by A itself. This asymmetry under time reversal may be the consequence of a special *initial* condition, similar to that characterizing irreversible master equations. In particular, if A is a conserved quantity, any back-flow, must be negligible. This condition represents a fact-like T-asymmetry that may be explained by assuming a sufficiently large and initially empty reservoir (comparable to the 'irrelevant channel' used in Sect. 3.2). If recurrence times are sufficiently large, the exponential law (4.44) may remain an excellent approximation, describing the decaying object for a very long time.

This disappearance of a 'substance' A from a given subsystem or region in space is an entirely classical model. However, the time dependence (4.44) is best known from radioactive decay in quantum theory, where A represents the non-decay *probability*. It is then regarded as the standard example of *quantum indeterminism* – usually understood as fundamental and law-like. This interpretation of (4.44) would mean that decay *events* occur at unpredictable though definite instants in time.

The decay law (4.44) defines an elementary master equation (3.48) with a Green's function $\hat{G}_{\text{ret}}$ simply given by the decay rate λ (see Sect. 4.1.2). Its foundation on time-symmetric fundamental dynamics (such as a universal Schrödinger equation) requires quite analogous assumptions, for example the negligibility of any back-flow into 'doorway states' that are directly coupled to A (see Fig. 3.4). Therefore, a conserved quantity has to disappear fast enough from such doorway states into 'deeper' (dynamically more distant) states, which must form a large reservoir.

A simple model is provided by the T-symmetric finite reaction chain

$$\frac{dA_n}{dt} = -(\lambda_n + \lambda_{n-1})A_n + \lambda_n A_{n+1} + \lambda_{n-1}A_{n-1} , \qquad (4.45)$$

with $n = 0, \ldots, N$, $\lambda_{-1} = \lambda_N = 0$, and the (improbable) initial condition $A_{n \neq 0} \approx 0$. $n = 1$ represents here the doorway channel of Sect. 3.2. For $\lambda_0 \ll \lambda_{n \neq 0}$, one obtains

$$\frac{dA_0}{dt} \approx -\lambda_0 A_0 , \qquad (4.46)$$

as long as $A_1 \ll A_0$. This requires only $\lambda t \ll N$, rather than $\lambda t \ll 1$, since all $A_{n \neq 0}$ will relax into partial equilibrium $A_{n \neq 0} \approx A_1$ on a short time scale (or just propagate away for $N \to \infty$).

Exponential decay can similarly be described by a deterministic wave equation on a continuum, where the small transition rate λ_0 is replaced by a potential barrier. It is irrelevant that the Schrödinger equation does here not describe the conserved quantity ('probability') itself. An *overall* time dependence according to a complex energy eigenvalue, $\psi(t) \propto \exp[-i(E_0 - i\gamma)t]$, would not be compatible with unitarity, but it may well represent an approximation that is valid in a bounded though growing spacetime region (Khalfin 1958, Petzold 1959, Peres 1980a) – similar to the reaction chain (4.45). Distant regions in space form a large reservoir.

In scattering theory, unstable states correspond to poles of the S-matrix $S_{nn'}(k)$, analytically continued into the complex plane, at points $k = k_1 - ik_2$ in the lower right half-plane ($k_1 > 0$ and $k_2 > 0$), where k is the wave number, $k^2 = k_1^2 - k_2^2 - 2ik_1k_2 = 2mE$. In the restricted spacetime region, where exponential behavior is observed after the incoming waves producing the decaying system have ceased, the wave function is dominated by the *Breit–Wigner* part (i.e., the pole contribution). This requires a (positive) *time delay* during the scattering process, which must be described by the relevant partial wave $\psi_l(r,t)Y_{lm}(\theta, \phi)$. Its radial factor $\psi_l(r,t)$ may be expanded in terms of energy eigenstates, $\psi_l^{(k)}(r,t) := \phi_{k,l}(r)e^{-i\omega(k)t}$, in the form

$$\psi_l(r,t) = \int_0^\infty f_l(k)\psi_l^{(k)}(r,t)dk$$

$$\xrightarrow[r \to \infty]{} \int_0^\infty f_l(k)\frac{e^{-ikr} - (-1)^l S_l(k)e^{ikr}}{r}e^{-i\omega(k)t}dk , \qquad (4.47)$$

where $S_l(k) = e^{2i\delta_l(k)}$ is the corresponding diagonal element of the S-matrix.

For sufficiently large values of t, the factor $e^{-i\omega(k)t}$ oscillates rapidly with k. This leads to destructive interference under the integral, except in regions of r and t where the phase $kr + \omega(k)t$ (for the incoming wave), or $kr - \omega(k)t + 2\delta_l(k)$ (for the outgoing one), is almost independent of k over the width of the wave packet $f_l(k)$ (which may be centered at k_0, say). For the outgoing wave, for example, this requirement means

$$\frac{d}{dk}\left[kr - \omega(k)t + 2\delta_l(k)\right]\Big|_{k_0} \approx 0 \quad \Longrightarrow \quad r \approx \frac{d\omega(k_0)}{dk_0}t - 2\frac{d\delta_l(k_0)}{dk_0} . \quad (4.48)$$

A noticeable delay compared to propagation with the group velocity $d\omega/dk$ requires a large value of $d\delta_l/dk$, such as in the vicinity of a complex pole of δ_l. For sufficiently large times t, but not too large distances r from the scattering center, and for initial momentum packets much wider than the size of the imaginary part k_2, only the pole contribution remains. For this one may write

$$S_l(k) = e^{2i\delta_l(k)} \approx \frac{k - k_1 - ik_2}{k - k_1 + ik_2} , \quad (4.49)$$

and hence $k_0 = k_1$ for the surviving wave packet that represents the decaying state. In this spacetime region, the contribution of the pole to (4.49) is given by its residue, whence

$$\psi_l(r,t) \xrightarrow[t \to \infty]{} -(-1)^l f_l(k_1) \int_0^\infty \frac{k - k_1 - ik_2}{k - k_1 + ik_2} \frac{e^{i\left[kr - \omega(k)t\right]}}{r} dk$$

$$\approx (-1)^l 2\pi k_2 f_l(k_1) \frac{e^{i\left[k_1 r - \omega(k_1)t\right]}}{r} \exp\left[k_2\left(r - \frac{d\omega(k_1)}{dk_1}t\right)\right] \quad (4.50)$$

(assuming $k_2 \ll |k_1|$). In the last factor one recognizes the 'imaginary part of the energy', $\gamma = k_2 d\omega(k_1)/dk_1$.

A positive delay (a 'retardation') of the scattered wave at the resonance requires

$$\frac{d\delta_l}{dk} \approx -\frac{d}{dk}\left(-\arctan\frac{k_2}{k - k_1}\right) = \frac{k_2}{(k - k_1)^2 + k_2^2} > 0 . \quad (4.51)$$

The pole must therefore reside in the lower half-plane. This condition is often referred to as *causality in scattering*, since the retardation specifies a direction in time related to intuitive causality (Chap. 2). This position of the poles is also used for deriving *dispersion relations* in T- or TCP-symmetric quantum field theory. However, no time direction can be specified by the structure of the S-matrix, since the latter is a consequence of the time-reversal-invariant Hamiltonian. Exponential decay is a fact-like asymmetry that would be reversed, using the *same* S-matrix, for scattering states with a time-reversed boundary condition. So one would have to force the outgoing wave rather than the incoming one to form a wave packet of limited duration. This would require the former decay products to arrive in the form of *advanced* coherent and exponentially growing waves over a very long span of time. A similar fact-like asymmetry, also caused by boundary conditions, occurred for the retarded radiation reaction of extended charges (2.31). The time arrow of exponential decay is thus determined by the boundary condition used in (4.47), rather than by the position of poles in the S-matrix.

The investigation of wave packets for non-relativistic particles decaying into free space beyond the pure pole contribution (Petzold 1959, Winter 1961, Peres 1980a) shows that deviations from the exponential law are essential not only at and just after formation of the decaying object, but also for very large times. In the extreme limit (when the decaying wave function has become unobservably small), exponential decay would be replaced by a power law. This consequence is in conflict with the interpretation of decay as a stochastic emission of particles (see Sect. 4.3.6). It must instead be understood as a *coherent back-flow* according to the dispersion of the outgoing wave. Although representing a very small effect in absolute terms, this deviation from exponential decay is even further reduced whenever the decay products interact with surrounding matter. In the usual case of strong coupling to the environment (such as absorption or decoherence), the exponential law remains an excellent approximation as long as the thermodynamical arrow characterizing absorbers remains valid. On the other hand, decay by emission of weakly interacting photons inside reflecting walls of a cavity has been confirmed to lead to the predicted deviations from exponential decay. It may even cause a 'coherent revival' of the decaying state (see Rempe, Walther and Klein 1987, Haroche and Kleppner 1989), but has also been observed in other situations (Wilkinson et al. 1997). Deviations from exponential decay depend crucially on the density of available final states. For example, coherent decay into a single final state is well known to lead to harmonic oscillation (complete periodic back-flow).

The Breit–Wigner contribution (4.50) describes a non-normalizable (even exponentially increasing) wave function. This result is an artifact of the pure pole approximation. The normalized state is correctly described by the wave packet $f_l(k)$ in (4.47), which has been replaced by a constant in (4.50). In an exact treatment, its square-integrable tails warrant normalizability even for large t by correcting the pure Breit–Wigner contribution.

In the general case, the scattering process would have to be described by a normalized time-dependent density matrix,

$$\rho_{lm,l'm'}(r,r';t) \xrightarrow[r\to\infty]{} \int_0^\infty \int_0^\infty \rho_{lm,l'm'}(k,k') \frac{e^{-ikr} - (-1)^l S_l(k)e^{ikr}}{r}$$

$$\times \frac{e^{ik'r'} - (-1)^{l'} S_{l'}^*(k')e^{-ik'r'}}{r'} e^{-i\left[\omega(k)-\omega(k')\right]t} dk\, dk' , \qquad (4.52)$$

where $\rho_{lm,l'm'}(k,k')$ is determined by the preparation procedure. After completion of the direct scattering process, and in the case of a resonance in the

l_0-wave, this leads approximately to

$$\rho_{lm,l'm'}(r,r';t) \xrightarrow[t\to\infty]{} \delta_{ll_0}\delta_{l'l_0}\rho_{l_0m,l_0m'}(k_1,k_1) \tag{4.53}$$

$$\times \int_0^\infty \int_0^\infty \frac{k-k_1-ik_2}{k-k_1+ik_2}\frac{k'-k_1+ik_2}{k'-k_1-ik_2}\frac{e^{i(kr-k'r')}}{rr'}e^{-i[\omega(k)-\omega(k')]t}dk\,dk' ,$$

in the relevant spacetime region. This approximation describes once again a *pure* Breit–Wigner wave packet (or at most a mixture of magnetic quantum numbers in the case of rotational symmetry).

Hence, there are no exactly exponential states (Gamow vectors) which would require or justify the 'rigging' of the Hilbert space of quantum mechanics. Similarly, there are no exact energy eigenstates in reality, since their infinite exponential tails according to $\exp(-\sqrt{-2mE}r)$ can never form completely within finite time. If exact energy eigenstates did occur by means of instantaneous quantum jumps, they would lead to superluminal effects (as has even been found surprising – see Hegerfeldt 1994). Even non-relativistically, stable and decaying states must form dynamically, that is, in accordance with a time-dependent Schrödinger equation, and hence with the time–energy uncertainty relation. Relativistically, an exactly bounded spatial support for a quantum state requires small (usually unobservable) uncertainties in energy and particle number, related to the Casimir effect or the Unruh radiation (see Sect. 5.2). Neglecting this consequence of quantum field theory (by arguing solely in terms of single-particle states, for example) leads to inconsistencies which illustrate the danger of remaining mathematically exact while not distinguishing between hypothetically fundamental and phenomenological (approximately valid) concepts – see also Sect. 4.3.2.

Radioactive decay is investigated in practice by means of *ensembles of many objects* in identical internal states (unstable nuclei, say). If these objects are distinguishable, for example by their position, their total state may be described as a direct product. According to the Schrödinger equation, each factor state will then evolve into a superposition of its initial state (with an exponentially decreasing amplitude) and a direct product of its final state and outgoing waves for the emitted objects. The *precise* time-dependent form of all components depends on what happens to the decay fragments (whether they are reflected, absorbed, or propagate into infinite space).

If the emitted particles (now assumed *not* to be reflected somewhere) are regarded as part of the environment of the decaying system, and thus traced out, the remaining N-atom density matrix describes a time-dependent *apparent ensemble* of different direct products. An ensemble variable $n(t) < N$ may here count the number of undecayed nuclei at each value of t (that is, in an ensemble of objects rather than potential states). The formal probabilities of these states in their apparent ensemble, $p_n(t)$, must then reflect the time-dependence of the Schrödinger amplitude, for example $p_0(t) = e^{-2N\gamma t}$. (In general, components with the same number n, but *different individual* decayed nuclei, also decohere from one another – see Sects. 4.3.5 and 4.3.6.)

This apparent ensemble of discrete numbers is dynamically approximately described by a master equation. Therefore, it is formally equivalent to an ensemble of solutions of a stochastic (Langevin-type) equation that essentially describes individual discrete 'histories' $n(t)$ – here in the form of 'descending staircase functions'. In terms of a universal Schrödinger equation, the number of undecayed nuclei n is a 'robust' property in the sense of Sect. 4.3.2 if decay can be assumed to be irreversible (in particular when monitored by detectors). The various dynamically robust branches of the wave function, arising by the fast but smooth action of decoherence, describe individual histories for integer numbers $n(t)$, which represent successions of almost discrete quantum jumps at certain times $t_1, t_2, \ldots$ (as discussed in Sect. 4.3.6). Similar staircase functions have now also been observed for decaying photons in a cavity (Gleyzes et al. 2006) – thus directly confirming Fig. 3.30 of Joos et al. (2003). However, deviations from exact steps can always be calculated if the interaction with the environment is known (Joos 1984): quantum theory is *not* a stochastic theory for quantum jumps.

4.6 The Time Arrow in Various Interpretations of Quantum Theory

> The truth could not be worth much
> if everybody was a bit right.

Physicists who completely agree about all applications of quantum mechanics often differ entirely about its interpretation, and even on the question of whether there remain *any* meaningful problems beyond the mere formalism (see Fuchs and Peres 2000). Although most of them would agree that quantum theory allows no more than probabilistic predictions, they often *derive* irreversible master equations, which describe an increase in entropy, from the deterministic and time-symmetric Schrödinger equation, using special initial conditions as in classical statistical physics (see Sect. 4.1.2). However, a dynamical probability interpretation must be relevant for the arrow of time – regardless of whether it is based on a fundamental stochastic (time-asymmetric) law or on an incompleteness of the theory (hidden variables) that refers to an unknown future. Its consequences cannot be avoided just by adding new words. For example, quantum theory is often called 'deterministic but acausal' – while this statement is then justified by the 'uncertainty' of classical properties (such as particle positions or momenta), which just *do not apply* to quantum states. Most physicists seem to disregard this consistency problem in an act of *Verdrängung*.

The deepest roots of these conceptual inconsistencies seem to arise from the fundamental difference between Heisenberg's and Schrödinger's 'pictures' (see Zeh 2004). While Heisenberg maintained classical concepts in principle (suggesting only a limitation of the 'certainty' of their values), Schrödinger described microscopic physical states by wave functions, which can be regarded as certain. The classical configuration space on which they are usually defined would thereby replace three-dimensional space as a new 'arena of dynamics' rather than describing *potential states*.[6] Whether wave functions or 'observables' (which *formally* replace the classical variables in the Heisenberg picture) carry the dynamical time-dependence is merely a *consequence* of the chosen picture.[7]

Although both pictures are equivalent when used to calculate formal expectation values for isolated systems, they describe the time arrow of quantum measurements in different ways. Most physicists seem to subscribe to one or the other picture (or perhaps a variant thereof) when it comes to interpretations ('probabilities for *what?*'). Typically, in the Schrödinger picture one regards the collapse of the wave function as a dynamical process, while in the Heisenberg picture it is viewed as an (extraphysical) increase of 'human knowledge'. I hope that keeping this difference in mind for the rest of this section may help to avoid some misunderstandings that often lead to emotional debate. One should therefore concentrate on what is actually *done* when the theory is successfully applied – though not in a merely pragmatic way. Which concepts are *fundamentally required*, rather than being approximately justified, or even mere tradition and prejudice?

Any meaningful concept of incomplete information or knowledge has to refer to an *ensemble* of possible states. For example, physical entropy, which quantifies irreversibility, is in quantum statistical mechanics defined by means of von Neumann's functional of the density matrix (4.4). According to Sect. 4.2, it measures the size of (genuine or apparent) ensembles of mutually orthogonal (hence operationally distinguishable) *wave functions*. While only *genuine* ensembles represent incomplete information, the time-dependence of the density matrix determines that of local entropy in general. Conservation of *global* von Neumann entropy reflects the unitarity of the von Neumann equation (when applicable) – equivalent to the unitarity *and* determinism of the Schrödinger equation. No ensemble of classical or any other (unknown)

---

[6] The identity of configuration space and space in single particle quantum mechanics is a consequence of the exceptional kinematics of mass points. This has led to a popular confusion of single-particle wave functions with spatial fields, and to the misnomer of a 'second quantization' in quantum field theory – see Zeh (2003).

[7] This contrast between the Heisenberg and the Schrödinger pictures has to be distinguished from the 'dualism' between two competing classical concepts (particles and fields) that is part of *one* (the Copenhagen) interpretation. In classical theory, particle positions and field strengths characterize *different physical objects*, which are both constituents of general physical systems. A dualism (or 'complementarity'), apparently required to characterize quantum objects, should more correctly be understood as a conceptual inconsistency, often attributed to a 'lacking microscopic reality'. However, this conceptual dualism applies only to the 'phenomenological reality' (see Sect. 4.3.2). A critical account of the origin of these conceptual problems can be found in Beller (1996, 1999).

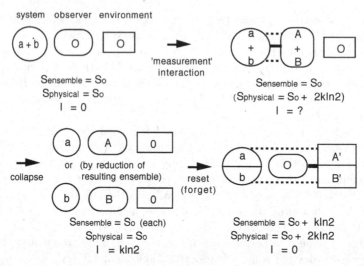

Fig. 4.3. Quantum measurement of a *superposition* $(|a\rangle + |b\rangle)/\sqrt{2}$ by means of a *collapse* process, here assumed to be triggered by the macroscopic pointer position. The initial entropy S_0 is smaller by one bit than in Fig. 3.5 (and may in principle vanish), since there is no initial ensemble 'a or b' for the property to be measured. Dashed lines before the collapse now represent quantum entanglement. (Compare the ensemble entropies with those of Fig. 3.5!) The collapse itself is often divided into two steps – see (4.54) below: first increasing the ensemble entropy by replacing the superposition by an ensemble, and then lowering it by reducing the ensemble (applying the 'or' – for macroscopic pointers only). The total increase of ensemble entropy, evident in the final diagram, is a consequence of the first step of the collapse. It brings the entropy up to its classical initial value of Fig. 3.5. The reset here illustrates also why decoherence is usually irreversible even when a measurement result is 'erased' (and even without a collapse – in which case the final ensemble entropy would again be S_0). From Chap. 2 of Joos et al. (2003)

variables representing the potential values of observables is 'counted' by von Neumann's entropy. Figure 3.5, characterizing classical measurements, *cannot* therefore be applied to quantum measurements. In terms of quantum states it has to be replaced by Fig. 4.3, which includes a collapse of the wave function. The transition from a superposition to an ensemble (depicted by the second step) affects the final value of von Neumann's 'ensemble' entropy (that would be reduced by a mere increase of information, as in the first step of Fig. 3.5). For similar reasons there can be no 'postselection' (no retarded increase of information about the past) by a quantum measurement, as suggested by Aharonov and Vaidman (1991): there is nothing to 'select' from in the absence of an ensemble of hidden variables.

A wave function *and* a set of classical configurations are kinematically used in Bohm's quantum theory (Bohm 1952, Bohm and Hiley 1993). This theory is often praised for exactly reproducing all predictions of conventional quantum theory in a deterministic way. However, this is not surprising, since it leaves Schrödinger's wave function entirely unchanged, while the assumed trajectories for classical states, which would *determine* all observed quantities according to this model, have to remain unobservable and in drastic conflict with classical intuition ('surrealistic') in order to reproduce the empirically confirmed quantum probabilities by means of their *postulated* statistical distribution. Because of the 'phenomenological' wave–particle dualism (see Sect. 4.3.2), it also remains controversial in this theory whether the classical configurations must contain photon positions or electromagnetic fields (Holland 1993).

Although wave functions and trajectories in configuration space are equally assumed to be real in this theory,[8] they are *treated* quite differently. While the former are usually regarded as 'given', the latter are always represented by an ensemble (without thereby contributing to the entropy). Their initial probability distribution in this ensemble, which has to be regarded as incomplete information, is *postulated* to comply with the Born rule. Since the Bohm trajectories themselves remain unobservable, they can be said to serve as no more than artificial and empirically unfounded selectors for the 'active' branch of the global wave function, to which the actual trajectory would be confined according to its dynamics. For example, entropy is calculated, in the form $S[\hat{P}|\psi\rangle\langle\psi|]$ with an appropriate Zwanzig projection $\hat{P}$, from such a *component* ψ – as though the wave function had been reduced by a real collapse (see Sect. 5 of Dürr, Goldstein and Zanghi 1993). While this description requires the same fact-like time asymmetry of the global wave function as decoherence, the selection of subsets of trajectories defines an *external* arrow of time. A justification of this different treatment of wave functions and Bohm trajectories is not at all obvious (see Zeh 1999b).

Similarly to Bohm's theory, collapse theories (Pearle 1976, Ghirardi, Rimini, and Weber 1986) and the Everett interpretation (Everett, 1957) also assume the wave function to represent a real physical object. This is in contrast to genuine hidden variables theories, which intend to *derive* or *explain* the wave function from some (hoped-for) more fundamental level of description. These latter theories are affected by various no-go theorems (such as Bell's theorem) if they are assumed to be local. Otherwise, however, it is hard to see what could be gained from them in comparison to the global wave function itself as a nonlocal object.

While collapse theories propose stochastic modifications of the Schrödinger equation, the Everett interpretation is based on the concept of 'splitting ob-

---

[8] "No one can understand this theory until he is willing to think of ψ as a real objective field rather than just a 'probability amplitude' " (Bell 1981). This statement may apply to quantum theory in general – quite in contrast to an analogous statement about Bohm's trajectories, which are empirically unfounded and thus have no more than 'religious' status. A stochastic theory can *always* be deterministically completed by means of unobservable variables: any by definition unobservable (pseudo-)random number generator will do.

servers' – a quite natural consequence in a kinematically nonlocal theory if observers have to remain local for dynamical reasons (see also Chap. 6). This splitting is facilitated by means of decoherence, defined as the dislocalization of superpositions (Sect. 4.3). Because of the locality of interactions, this process describes an effective dynamical decoupling (called a 'branching') of the global wave function into components that are characterized by different quasi-classical (robust and quasi-local) properties – including those of systems that can be regarded as observers. This decoherence has turned out to be the most efficient and most ubiquitous irreversible process in Nature.

While decoherence eliminates the basic motivation for the Heisenberg–Bohr interpretation (that *presumes* genuine classical concepts, and their values as 'coming into being' during fundamental irreversible events outside the laws of physics – see Pauli's remark towards the end of the Introduction), this aspect of decoherence may not have been duly appreciated even by some authors who significantly contributed to it (Omnès 1988, 1992, Gell-Mann and Hartle 1990, 1993 – see also Omnès 1998). Apparently guided by the Heisenberg picture, they investigate consequences of decoherence for certain 'consistency conditions' (originally proposed by Griffiths 1984), which are assumed to regulate the applicability (or different kinds of 'truth') of varying classical concepts within variable presumed uncertainties ('coarse graining') at selected discrete times. However, environmental decoherence allows one to *derive* quasi-classical concepts in terms of *wave packets* (all that is needed) – close to what Schrödinger had originally in mind. Their apparent ensembles, formally described by the density matrix, obey master equations – such as (4.35), and *in this way consistently define* quasi-classical 'histories'.

In contrast to Bohm's or Everett's theories, consistent histories usually presume an *absolute* quantum arrow (see Hartle 1998 and footnote 2 of Chap. 5) – just as collapse models do. Their selection by a formal 'consistency' requirement may indeed be inconsistent itself (Kent 1997), while master equations derived by means of decoherence are never assumed to hold exactly. Consistent histories have also been claimed to be equivalent to stochastic trajectories *of wave functions*, in the 'quantum state diffusion model' assumed to exist for *all* systems (Diósi et al. 1995, Brun 2000 – see Sect. 4.4). However, this would be in conflict with the empirically confirmed quantum nonlocality.

On the other hand, the concept of apparent ensembles of wave functions in decoherence theory, or the density matrix in general, are based on a probability interpretation (as explained in Sect. 4.2), while the question 'probabilities for (or information about) *what*?' has not yet been answered on a fundamental level. Many quantum phenomena seem to favor the answer: new wave functions (such as narrow wave packets). For example, a spot on the photographic plate (regarded as a measurement 'pointer') has to be described in terms of local molecular *wave functions* – not in terms of any classical variables. The strongest support yet for such an interpretation may come from an analysis of the decoherence of neuronal states in the brain (Tegmark 2000, see also Zeh 2000). These quasi-classical neuronal states may form the final link (or the ultimate pointer basis) in a chain of observational interactions that are all describable in terms of a global wave function.

The superposition of such quasi-classical observer states, which would result like Schrödinger cats from the unitary dynamics, was evidently the reason for postulating a collapse of the wave function as a *real* dynamical process (von Neumann's 'first intervention').[9] It may be formulated in two steps (see Fig. 4.3):

$$|\psi\rangle\langle\psi| = \underbrace{\sum_{mn} |\phi_m\rangle c_m c_n^* \langle\phi_n|}_{} \longrightarrow \underbrace{\sum_n |\phi_n\rangle |c_n|^2 \langle\phi_n|}_{} \longrightarrow \underbrace{|\phi_{n_0}\rangle\langle\phi_{n_0}|}_{},$$

$$\text{with} \qquad S = 0 \qquad \longrightarrow \qquad S \geq 0 \qquad \longrightarrow \qquad S = 0 \,. \tag{4.54}$$

They represent (1) the transition from a pure state into an ensemble (characterized by an increase in von Neumann's ensemble entropy), and (2) the selection of a specific state from this ensemble (thus lowering the ensemble entropy as depicted by the first step of Fig. 3.5). The first step can be represented by a master equation that describes a loss of information, while the complete stochastic process (4.54) corresponds to a quantum Langevin equation (a Langevin equation for wave functions – see Sect. 4.4). Since the complete process describes an individual physical evolution, it has to be used when calculating the changing *physical* entropy according to (3.58). The master equation describes the dynamics of an ensemble that represents entanglement *and* lacking information about the stochastically evolving wave function.

If macroscopic properties α, say, are again regarded as 'always given' (as in Sect. 3.3.1), *physical entropy* can be characterized by a function $S(\alpha) = k \ln N_\alpha$, similar to the last term of (3.58), where N_α is now the dimension of the subspace representing a fixed value (or small interval) of α. The time dependence of this entropy,

$$S(t) = S(\alpha(t)) \,, \tag{4.55}$$

is then determined by the macroscopic dynamics $\alpha(t)$, which in general includes a succession of collapse 'events', each one a dynamical projection onto a subspace corresponding to definite macroscopic properties. This means that *the stochastic collapse is part of the macroscopic dynamics* $\alpha(t)$. It is dynamically objectivized by the process of decoherence that describes an *apparent* collapse. In contrast to classical Hamiltonian dynamics, which determines the time dependence of any macroscopic quantity, $\alpha(t) := \alpha(p(t), q(t))$ – see Sect. 3.3.1, the Schrödinger equation does *not* determine $\alpha(t)$. Therefore, the

---

[9] Despite frequent claims to the contrary, the dynamical collapse was never part of the Copenhagen interpretation, although quantum jumps have traditionally been used as an argument against a wave function representing reality. The collapse is then claimed to represent 'just a normal increase of information', even though an ensemble representing such incomplete information is excluded.

dynamics (4.32) cannot describe the transformation of entropy into lacking information in accordance with the negentropy principle (3.62). In other words, not even *macroscopically* different states have to possess different predecessors ('sufficient reasons') in quantum theory. Such causal predecessors, if they existed, would have to be counted by the initial ensemble entropy as in Fig. 3.5.

Would the collapse, if *used* in this way as part of the dynamics of wave functions, now specify an arrow of time that could perhaps even be responsible for irreversible thermodynamics? The ensemble entropy (4.4) would increase only during the auxiliary first step of the collapse (describing ignorance about the outcome). For entangled systems, for example those occurring after an interaction of type (4.32), an individual ('real') collapse,

$$\psi = \sum_n c_n \phi_n \Phi_n \xrightarrow{t} \phi_{n_0} \Phi_{n_0} , \qquad (4.56)$$

does *not* alter the ensemble entropy. However, it specifies an arrow as it transforms the entangled state into a factorizing one. Therefore, the additive ('physical') entropy *decreases* in this process after it may have correspondingly *increased* during the interaction (4.32), since

$$S[\hat{P}_{\text{sep}}|\psi\rangle\langle\psi|] \geq S[\hat{P}_{\text{sep}}|\phi_{n_0}\Phi_{n_0}\rangle\langle\phi_{n_0}\Phi_{n_0}|] = 0 . \qquad (4.57)$$

A collapse has never been confirmed empirically as a dynamical process. It has nonetheless to be *taken into account* regardless of its interpretation before (or, at least, when) the observer becomes aware of the macroscopic pointer position. As he thereby becomes himself quantum correlated with the pointer state Φ_{n_0}, the corresponding 'state-of-being-conscious' in his brain also becomes a pure state as a consequence (if not the origin) of this collapse. This 'observer state' (whatever it may be in detail) can thus be used for postulating a psycho-physical parallelism in accordance with von Neumann's intentions (see also London and Bauer 1939). There is thus no need for genuine classical (or other) variables *anywhere in-between* the observed microscopic system and the brain of the observer. Spontaneous localization of the pointer position in an apparatus would *not* lead to a pure state-of-being-conscious. Instead, one could regard the neuronal system as the true measurement device, while treating the whole outside world as one quantum system. Although this description of observations by conscious beings remains vague in detail, it is all that is *in principle* required for a physical formulation in terms of quantum states. While a 'real' collapse is usually assumed to occur as soon as certain phase relations have become irreversibly dislocalized, this assumption is merely convenient, as it comes close to reproducing a classical description.

Does the entropy-reducing collapse, wherever it may occur (or simply be *applied*), then have to be regarded as a quantum mechanical revival of Maxwell's demon, which could classically be exorcized in Sect. 3.3.2? Lubkin (1987) demonstrated that, just as in the classical case, this entropy decrease according to the collapse cannot be utilized in a cyclic process that would allow the construction of a perpetuum mobile of the second kind. However, the

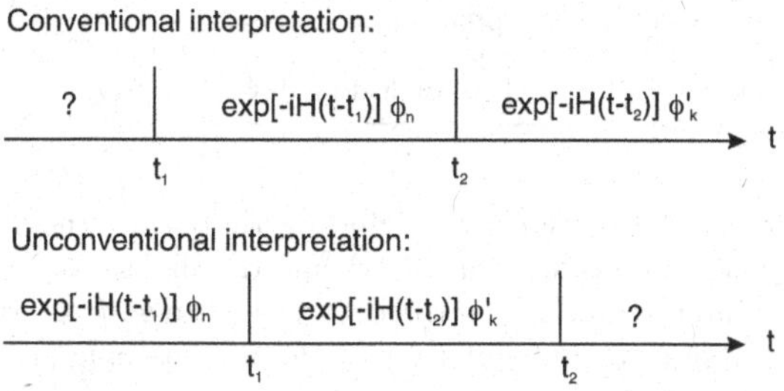

Conventional interpretation:

?	$\exp[-iH(t-t_1)]\,\phi_n$	$\exp[-iH(t-t_2)]\,\phi'_k$

t_1 $\qquad\qquad$ t_2 $\qquad\qquad\qquad$ t

Unconventional interpretation:

$\exp[-iH(t-t_1)]\,\phi_n$	$\exp[-iH(t-t_2)]\,\phi'_k$	?

t_1 $\qquad\qquad\qquad$ t_2 $\qquad\qquad\qquad$ t

Fig. 4.4. Dynamics of the wave function in the case of retarded (conventional) and advanced (acausal) collapse. In contrast to classical waves, the choice of the usual ('retarded') interpretation of the collapse is a matter of pure convention

collapse may have important entropy-reducing consequences in a non-cyclic cosmic evolution (see Sect. 6.1).

It appears furthermore noteworthy that the usual form of the collapse is based on an extension of *intuitive causality* (Chap. 2) to a region where by assumption it cannot be confirmed. For example, the expression $|\langle\phi'_k|\phi_n(\Delta t)\rangle|^2$, with $\phi_n(\Delta t) := \exp(-iH\Delta t)\phi_n$, defines the probability of finding the state ϕ'_k in an appropriate measurement at time $t_2 = t_1 + \Delta t$, provided the system was found in the state ϕ_n in a previous measurement (of the first kind) at time t_1. This interpretation assumes the wave function to collapse into the state ϕ_n during the first measurement, and then to evolve unitarily according to the Hamiltonian H until it is measured again. An equivalent (though unconventional) time-reversed interpretation may be obtained from the identity of the above matrix element with $\langle\phi'_k(-\Delta t)|\phi_n\rangle$. This means that the wave function may as well be assumed to collapse during the first measurement in an 'acausal' manner from an 'advanced' state ϕ_n into $\phi'_k(-\Delta t)$, which will then unitarily evolve into the state $\phi'_k = \phi'_k(0)$ just *before* the second measurement starts (Penrose 1979). These two versions of the collapse are indicated in Fig. 4.4. The second one is counterintuitive, since the observed system would have to 'know in advance' what kind of measurement will be performed, and when. Its exclusion is therefore an application of *intuitive causality*, similar to the exclusion of advanced fields, while the equivalence of the two descriptions is based on the T-symmetry of the formal quantum probabilities (Aharonov, Bergmann and Lebowitz 1964).

In contrast to the advanced electromagnetic fields (Sect. 2.4), which can be excluded empirically by means of small test charges, our preference for the causal version of the collapse is purely conventional. However, macroscopic registration devices, described by states Φ, that are continuously monitored by the environment in accordance with the time arrow of increasing entanglement or quantum causality, have to be assumed to possess quasi-classical states at all times. In this case, their states during two successive measurements must be described by $\Phi_0(t)$ for $t < t_1$, by $\Phi_n(t)$ for $t_1 < t < t_2$, and by $\Phi_{nk}(t)$ for $t > t_2$, where the number of indices corresponds to the increasing number of

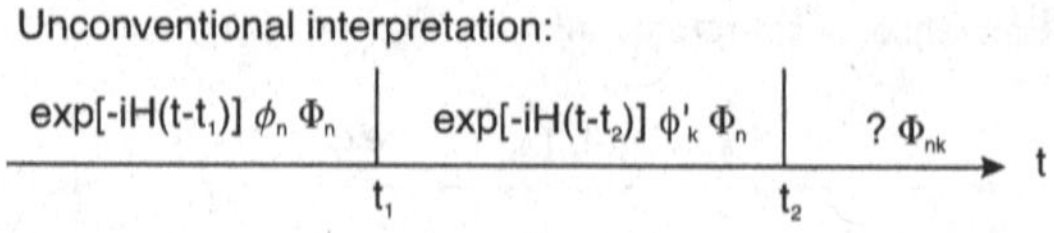

Unconventional interpretation:

$$\underrightarrow{\left.\exp[-iH(t-t_1)]\ \phi_n\ \Phi_n\ \right|\ \left.\exp[-iH(t-t_2)]\ \phi'_k\ \Phi_n\ \right|\ \ ?\ \Phi_{nk}}_{\displaystyle t_1 \qquad\qquad\qquad t_2}\ t$$

Fig. 4.5. Behavior of the total state (including that of the measurement device) under the unconventional assumption of advanced collapse and 'continuous measurement' of the pointer position, described by the state Φ. Only the wave function of the microscopic system ϕ, but not that of the macroscopic pointer, depends on the convention

registered (robust) measurement results (see also Bläsi and Hardy 1995). This holds even in the acausal collapse version – as indicated in Fig. 4.5.

All these ambiguities can be avoided if the Schrödinger equation is not modified at all. For a wave function that is assumed to describe reality completely, this leads to the Everett interpretation. There is then no law-like quantum arrow of time – but how can measurements *lead to* any definite and irreversible results?

Everett's interpretation is based on the Schrödinger picture: "This paper proposes to regard pure wave mechanics as a complete theory" (Everett 1957, see also Zeh 1970, 1973). So *there are no longer any observables* to be introduced as further fundamental ingredients of the theory, as required in the Heisenberg picture. (The pragmatic equivalence of both pictures is in any case questionable for open, that is, locally non-Hamiltonian systems.) In particular, the 'many-worlds interpretation', which assumes many simultaneous histories formed by *classical* properties (DeWitt 1971, Deutsch 1997), or an ensemble of Feynman paths in classical configuration space (Sokolovsky 1998), misinterprets Everett's proposal by presuming classical concepts. A *superposition* of Feynman paths would be identical to a time-dependent wave function as an individual dynamical state.

The ultimate observer states (χ_n^{obs}, say), which differ in general in separate 'branches' of the wave function, need *not* necessarily represent quasi-classical states. Everett concluded from the Schrödinger equation that all components of (4.32) continue to exist in *one* superposition, $\sum \psi^{(n)} \chi_n^{obs}$ ("All components are actual"). His point is that they can be *experienced* only separately because of their separate observer states χ_n^{obs}, which are here *postulated* on empirical grounds to represent subjective awareness. Evidently, no local observer state is defined in the *global* Everett wave function. This interpretation answers von Neumann's quest for a psycho-physical parallelism in quantum mechanical terms without introducing a collapse (see Zeh 1970, 1979, 2000, Squires 1990, Lockwood 1996).

Since configuration space assumes the role of space as the arena of reality in all versions of the Schrödinger picture, the *fork of indeterminism* that seems to characterize probabilistic quantum theory is reinterpreted in the Everett interpretation as a *fork of causality* (see footnote 1 of Chap. 2). All observer states χ_n^{obs} (with different n) must be remembering the same *pre*-measurement history. To all of them, the rest of the world is thereafter described by their different 'relative states' $\psi^{(n)}$ (their co-factor states, which these observers may renormalize for convenience). The formal 'plus' of a superposition is objectively reduced to an effective 'and' by the irreversible dislocalization of superpositions, while the 'or' is observer-related ('subjective') – though objectivized with respect to correlated components of different observers.

Only if both factors of all components of the Everett wave function were defined to be mutually orthogonal would the sum of their products, $\sum \psi^{(n)} \chi_n^{obs}$, resemble the Schmidt representation (4.27) (Kübler and Zeh 1973, Zeh 1973, Albrecht 1992, 1993). This representation depends furthermore on the precise borderline between observer and the rest of the world. However, the essential (and objective) aspect of decoherence is the irreversible dynamical spreading of a superposition over *many* local subsystems $\Phi^{(k)}$ (possibly including observers) in the form $\sum_n c_n \phi_n \prod_k \Phi_n^{(k)}$ – thereby propagating in accordance with the relativistic spacetime structure.

As 'the other' robust components which must form according to the Schödinger equation are not observable to 'us', the assumption of their existence is *operationally* meaningless (see Sect. 4.3.2 about the concept of 'operational reality'). The Everett interpretation is therefore indistinguishable from an appropriately chosen collapse interpretation. In the Everett interpretation this equivalence requires the additional assumption that components only *branch* with increasing time, but in practice never (re)combine according to

$$\sum_n c_n \phi_n \Phi_n \quad \xrightarrow[t]{} \quad \left(\sum_n c_n \phi_n \right) \Phi_0 \, . \tag{4.58}$$

Because of the T-symmetric Schrödinger dynamics, the absence of this process requires quite generally that no suitable (conspiratorial) components $n \neq n_0$ *exist* on the LHS according to the initial condition for the global wave function. This demonstrates also that the wave function cannot merely describe *possibilities* (in contrast to wave functions that *could* have arisen in a stochastic process, for example). The law-like arrow of the collapse is thus replaced by a fact-like arrow. This 'quantum causality' is formally analogous to Boltzmann's *Stoßzahlansatz*, which requires that correlations are irrelevant in practice after being formed in collisions.

One may postulate an appropriate cosmic initial condition for this purpose by requiring that *all* existing nonlocal quantum correlations, such as those on the LHS of (4.58), are retarded, that is, have been *caused* somewhere during the past history of the quantum Universe. At the big bang ($t = 0$, say) one would thus have to assume a completely uncorrelated state, that is, a wave function of a form like

$$\psi(t \to +0) = \lim_{\Delta V_k \to 0} \prod_k \psi_{\Delta V_k}^0 \, , \tag{4.59}$$

where $\psi_{\Delta V_k}$ are local states on appropriate small volume elements ΔV_k. This kind of state is invariant under the corresponding Zwanzig projection $\hat{P}_{\text{local}}$ by definition. However, the assumption (4.59) appears 'natural' only to our causal prejudice (thus applying the *double standard* in Price's terminology). In contrast to its classical counterpart of initially absent (or future-irrelevant) correlations, this condition is not of a *statistical* nature (meaningful only for an ensemble that describes incomplete information), but a condition on the objective state of the quantum Universe. It would explain both the absence of quantum recoherence *and* the applicability of thermodynamical master equations.

Even though the Everett branches are sufficiently defined by means of decoherence (that is, by the dislocalization of superpositions – Sect. 4.3), this does not yet define any probabilities for them, as they are required by the Born rules. The branches are *all* assumed to exist once (with different albeit as yet physically meaningless norms). Everett's original claim that the probability interpretation is a consequence of the formalism was based on the density matrix, and so must be circular (see Sect. 4.1.2). Graham (1970) therefore considered *series* of N equivalent measurements (N subsequent branchings – similar to series of decay events discussed at the end of Sect. 4.5). He was then able to demonstrate that the total norm of all those of the resulting Everett branches which represent series of outcomes that differ significantly from the Born rule must become extremely small, and vanish in the limit $N \to \infty$ (see also Jammer 1974). While this result permits an elegant formulation of the probability postulate (by assuming merely that we happen to live in an Everett branch of not extremely small norm), this assumption is still completely equivalent to what is to be derived.[10]

Therefore, the probability measure has to be regarded as an *empirical* input to the theory (just like the Schrödinger dynamics, for example). The norm is a plausible candidate for this measure, because it is dynamically conserved under the Schrödinger equation. While stochastic collapse models postulate quantum probabilities as part of their dynamics, the Everett interpretation must *equivalently* assume that 'we' are living in a branch that has been selected by chance in accordance with the Hilbert space norm.

General Literature: Jammer 1974, Busch, Lahti and Mittelstaedt 1991, d'Espagnat 1995, Schloßhauer 2004, Zeh 2005.

---

[10] This may be illustrated by the example of results obtained for N subsequent measurements distinguishing between spin states $|+\rangle$ and $|-\rangle$ in N identical initial superpositions $a|+\rangle + b|-\rangle$. Since the *number* of branches which contain n 'spin-ups', say, is then statistically given by the binomial coefficient $\binom{N}{n}$ regardless of the values of the coefficients a and b, the distribution of measurement outcomes n over many such series of N measurements, $p_N(n)$, would form a Poisson distribution centered at the required value if and only if the probability *for each branch* is assumed to be given by its norm $|a|^{2n}|b|^{2(N-n)}$. This is precisely Born's probability rule. (Example provided by Erich Joos.)

5

The Time Arrow of Spacetime Geometry

In the framework of general relativity, gravity is a consequence of spacetime curvature. Its dynamical laws (Einstein's field equations) are again symmetric under time reversal. However, if their *actual* global solution, that is, the observed spacetime, is asymmetric (such as a forever expanding universe), this must affect the dynamics of all matter. While this was well known, it came as a surprise during the early 1970s that strongly gravitating systems possess thermodynamical properties, thus indicating an intimate connection between two seemingly very different fields of physics.

Gravitating systems are already thermodynamically peculiar in Newton's theory, since they possess negative heat capacity, resulting from the universal attractivity of this force. In particular, attractive forces which depend homogeneously on the minus second power of distance, such as gravity and Coulomb forces, lead according to the virial theorem to the relation

$$\overline{E_{\text{kin}}} = -\frac{1}{2}\overline{E_{\text{pot}}} = -E \, , \tag{5.1}$$

between the mean values of kinetic and potential energies, and therefore between them and the total energy. This virial theorem is valid for mean values over a (quasi-)period of the motion, or approximately (in the case of semi-stable states) for mean values defined over sufficiently large intervals of time. In quantum theory, mean values have to be replaced by expectation values on proper (normalizable) energy eigenstates. The theorem can then be conveniently proved using Fock's *ansatz* $\psi(\lambda r_1, \ldots, \lambda r_N)$ and the homogeneity of T and V in a variational procedure, $\delta(\langle\psi|T + V|\psi\rangle/\langle\psi|\psi\rangle) = 0$, with respect to λ. So it must also hold for expectation values on density matrices whose non-diagonal elements can be neglected in the energy basis. (For relativistic generalizations of the virial theorem see Gourgoulkon and Bonazzola 1994.)

The anti-intuitive negative sign relating kinetic and total energy in (5.1) means, for example, that satellites are *accelerated* by friction when they enter the earth's atmosphere, and that stars *heat up* by radiating energy away. This second example is valid only as far as the quantum mechanical zero-point energy does not dominate $\overline{E_{\text{kin}}} = \text{Trace}\{\rho T\}$ – as it would in white dwarf stars or solid bodies. Early astrophysicists believed instead that stars always cool down in the course of time. The virial theorem also means that the heat flow from hot to cold objects which are governed by gravity causes a thermal inhomogeneity to *grow*.

To construct an example, first consider a monatomic ideal gas in two vessels under different conditions, but under exchange of energy (heat), $\delta U_1 = -\delta U_2$, and particles, $\delta N_1 = -\delta N_2$. Their partial entropies according to (3.14) are given by

$$S_i = kN_i \left(\frac{3}{2} \ln T_i - \ln \rho_i + C \right) , \tag{5.2}$$

with $i = 1,2$ distinguishing the two vessels. Since the internal energy, $U = \overline{E_{\text{kin}}}$, is here $U = (3/2)NkT$, the total change of entropy becomes for fixed volumes V_i, or for fixed densities $\rho_i = N_i/V_i$,

$$\delta S_{\text{total}} = \delta S_1 + \delta S_2 = \left(\frac{1}{T_1} - \frac{1}{T_2} \right) \delta U_1 + k \left(\frac{3}{2} \ln \frac{T_1}{T_2} - \ln \frac{\rho_1}{\rho_2} \right) \delta N_1 . \tag{5.3}$$

This expression describes entropy changes δS_1 and δS_2 with opposite signs, which cancel only in thermodynamical equilibrium ($T_1 = T_2$ and $\rho_1 = \rho_2$). In this situation without gravity, an entropy increase in accordance with the Second Law requires a *reduction* of thermal and density inhomogeneities (except for the transient *thermo-mechanical effect*, that is, a thermally induced pressure difference that is caused by the temperature dependence of the second term).

However, the density of a gravitating star is not a free variable that can be kept fixed (as in the laboratory). A typical star, assumed for simplicity to be in thermal equilibrium, may to a very good approximation also be described as an ideal gas. Its temperature and volume are then related by means of the virial theorem according to

$$NT \propto U = \overline{E_{\text{kin}}} \propto -\overline{E_{\text{pot}}} \propto \frac{N^2}{R} \propto \frac{N^2}{V^{1/3}} , \tag{5.4}$$

that is, $V \propto N^3/T^3$. The entropy (5.2) of a star is therefore

$$S_{\text{star}} = kN \left(\frac{3}{2} \ln T - \ln N + \ln V + C \right)$$

$$= kN \left(-\frac{3}{2} \ln T + 2 \ln N + C' \right) . \tag{5.5}$$

In the second line, the signs of $\ln T$ and $\ln N$ are reversed. The total entropy change of a star embedded in an interstellar gas, $\delta S_{\text{star}} + \delta S_{\text{gas}}$, becomes after again using the virial theorem in the form $E_{\text{star}} = -U_{\text{star}}$,

$$\delta S_{\text{total}} = \left(\frac{1}{T_{\text{star}}} - \frac{1}{T_{\text{gas}}} \right) \delta E_{\text{star}} + k \ln \left[\frac{C'' N_{\text{star}}^2 \rho_{\text{gas}}}{(T_{\text{star}} T_{\text{gas}})^{3/2}} \right] \delta N_{\text{star}} . \tag{5.6}$$

While heat must still flow from the hot star into cold interstellar space in order to comply with the Second Law, this leads now to a further increase of the star's temperature, and the accretion of matter – provided the 'star' is already sufficiently massive. Thermal and density inhomogeneities thus *grow* in the generic astrophysical situation, although there are also 'pathological' objects with non-periodic motion, such as gravitationally collapsing spherical matter shells or pressure-free dust spheres, for which the virial theorem does not hold.

These arguments show that the evolution of normal stars is dynamically controlled by thermodynamics rather than by gravity itself. If the thermodynamical arrow of time did change direction in a recontracting universe (as suggested by Gold 1962 – see Sect. 5.3), stars and other gravitating objects would have to re-expand by means of advanced incoming radiation in spite of their attractive forces.

A homogeneous universe must therefore describe an unstable state of very low entropy (though a 'simple' state in the sense of Sect. 3.5). So one may ask whether the evolution of matter into inhomogeneous clumps under gravitational forces represents an entropy capacity that is sufficient to explain the observed global thermodynamical arrow of time. The apparently required *Kaltgeburt* of the Universe might then be replaced by a *homogeneous birth*, since inhomogeneous local contraction leads to the formation of strong temperature and density gradients.

In order to estimate the improbability (negentropy) of a homogeneous universe, one has to know the maximum entropy that can be gained by gravitational contraction. Conceivable limits of contraction are:

- *Quantum degeneracy* (primarily of electrons) is essential for the stability of solid gravitating bodies and white dwarf stars. By emitting heat, these objects cool down rather than further heating up.
- *Repulsive short range forces* are important in neutron stars, for example.
- *Gravitation* itself may lead to black holes even in Newton's theory. Any radiation with bounded velocity cannot escape from the surface of a sufficiently dense and massive object. If this velocity bound is as universal as gravity (as in the theory of relativity), the further fate of matter inside this critical surface remains completely *irrelevant* to an external observer. This surface defines an *event horizon* for him. Matter disappearing behind the horizon is irreversibly lost except for its long range forces, such as gravity itself. In particular, it can no longer participate in the thermodynamics of the Universe.

Such *non-relativistic black holes* were discussed by Laplace as early as 1795, and before him by J. Mitchel at Cambridge. In general relativity, black holes are described by specific spacetime structures. This leads to the further consequence that neither of the first two mentioned limits to gravitational contraction may prevent an object of sufficiently large mass (that could always be reached by further accretion of matter) from collapsing into a black hole. Repulsive forces would give rise to a positive potential energy, that must eventually dominate as a source of gravity, while the increasing zero point pressure of a degenerate Fermi gas would force the fermions into effective bosons that may form a further contracting condensate.

Therefore, only black holes define a realistic upper limit for entropy production by gravitational contraction of matter from the point of view of an external observer. But what is the value of the entropy of a black hole? This question cannot be answered by investigating relativistic stars, that is, equilibrium systems, since the essential stages of the collapse proceed irreversibly. However, a unique and finite answer is obtained from a quantum aspect of black holes, viz., their Hawking radiation (Sect. 5.1).

Since in general relativity the spatial curvature represents a dynamical state (see Sect. 5.4), it may itself carry entropy. Its dynamics is described by Einstein's field equations

$$G_{\mu\nu} = 8\pi T_{\mu\nu} , \tag{5.7}$$

in units with $G = c = 1$, where $T_{\mu\nu}$ is the energy–momentum tensor of matter. They define an initial (or final) value problem, since they are essentially of hyperbolic type (see Sect. 2.1). The Einstein tensor $G_{\mu\nu}$ is a linear combination of the components of the Ricci tensor $R_{\mu\nu} := R^{\lambda}{}_{\mu\lambda\nu}$, that is, the trace of the Riemann curvature tensor. Forming this trace is analogous to forming the d'Alembertian in the wave equation (2.1) for the electromagnetic potential from its matrix of second derivatives $\partial_\nu\partial_\lambda A^\mu$. Aside from nonlinearities (that are responsible for the self-interaction of gravity), the Riemann curvature tensor is similarly defined by the second derivatives of the metric $g_{\mu\nu}$, which thus assumes the role of the gravitational potential (analogous to A^μ in electrodynamics). In both cases, the trace of the tensor of derivatives is determined locally by the sources, while its trace-free parts represent the degrees of freedom of the vector or tensor field, respectively, which can therefore be freely *chosen* initially (as an incoming field).

Penrose (1969, 1981) used this freedom to conjecture that the trace-free part of the curvature tensor (the *Weyl tensor*) vanished when the Universe began. This situation describes a 'vacuum state of gravity', that is, a state of minimum gravitational entropy, and a space as flat as is compatible with the sources. It is analogous to the cosmic initial condition $A^\mu_{\text{in}} = 0$ for the electromagnetic field discussed in Sect. 2.2 (with Gauss's law as a similar constraint). Gravity would then represent a retarded field, requiring 'causes' in the form of advanced sources. Since Penrose intends to explain the thermodynamical arrow, too, from this initial condition (see Sect. 5.3), his conjecture revives Ritz's position in his controversy with Einstein (see Chap. 2) by applying it to gravity rather than to electrodynamics.

In the big bang scenario, the beginning of the Universe is characterized by a past time-like curvature singularity (where time itself began). Penrose used this fact to postulate his Weyl tensor hypothesis on all past singularities, since this would allow only *one* of them: a uniform big bang. In the absence of an *absolute* direction of time, the past would then be distinguished from the future precisely and solely by this asymmetric boundary condition and its consequences (again introducing a 'double standard'). If the Weyl tensor condition could be derived from some other assumptions that did *not* arbitrarily select a time direction, it would have to exclude inhomogeneous future singularities as well. This may again lead to dynamical consistency problems, but it would not rule out collapsing objects to *appear* as black holes to external observers (see Sects. 5.1 and 6.2.3).

5.1 Thermodynamics of Black Holes

In order to discuss the spacetime geometry of black holes, it is convenient to consider the static and spherically symmetric vacuum solution, discovered by Schwarzschild and originally expected to represent a point mass. In terms of spherical spatial coordinates, this solution is described by the metric

$$ds^2 = -\left(1 - \frac{2M}{r}\right)dt^2 + \left(1 - \frac{2M}{r}\right)^{-1}dr^2 + r^2(d\theta^2 + \sin^2\theta\,d\phi^2)\,. \quad (5.8)$$

Here, r measures the size of a two-dimensional sphere – though *not* the distance from $r = 0$. This metric form is singular at $r = 0$ and $r = 2M$, but the second singularity, at the *Schwarzschild radius* $r = 2M$, is merely the result of an inappropriate choice of these coordinates. The condition $r = 2M$ describes a surface of fixed area $A = 4\pi(2M)^2$ (using Planck units $G = c = \hbar = k_B = 1$) in spite of moving outwards at speed of light. In its interior (that is, for $r < 2M$) one has $g_{tt} = 2M/r - 1 > 0$ and $g_{rr} = (1 - 2M/r)^{-1} < 0$. Therefore, r and t interchange their physical meaning as spatial and temporal coordinates. This internal solution is *not* static, while the genuine singularity at $r = 0$ represents a time-like singular boundary rather than the space point expected by Schwarzschild.

Physical (time-like or light-like) world lines, that is, curves with $ds^2 \leq 0$, hence with $(dr/dt)^2 \leq (1 - 2M/r)^2 \to 0$ for $r \to 2M$, can only approach the Schwarzschild radius parallel to the t-axis (see Fig. 5.1). Therefore, the interior region $r < 2M$ is physically accessible only via $t \to +\infty$ or $t \to -\infty$, albeit within finite proper time. These world lines can be extended regularly into the interior when t goes beyond $\pm\infty$. Their proper times continue into the physically finite future (for $t > +\infty$) or past (for $t < -\infty$) with the new time coordinate $r < 2M$. There are therefore *two* internal regions (II and IV in the figure), with their own singularities at $r = 0$ (at a finite distance in proper times). These internal regions must in turn each have access to a new *external*

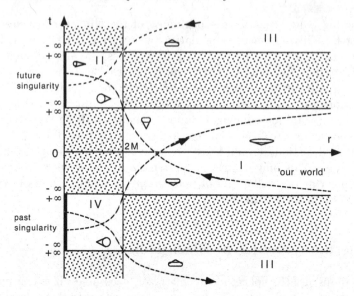

Fig. 5.1. Extension of the Schwarzschild solution from 'our world' beyond the *two* coordinate singularities at $r = 2M$, $t = \pm\infty$. Each point in the diagram represents a 2-sphere of size $4\pi r^2$. A consistent orientation of forward light cones (required from the continuation of physical orbits, such as those represented by *dashed lines*) is indicated in the different regions. There are also *two* genuine curvature singularities with coordinate values $r = 0$

region, also in their past or future, respectively, via different Schwarzschild surfaces at $r = 2M$, but with opposite signs of $t = \pm\infty$. There, proper times have to *decrease* with growing t. These two new external regions may then be identified with one another in the simplest possible topology (region III appearing twice in the figure).

This complete Schwarzschild geometry may be described by means of the regular *Kruskal–Szekeres coordinates* u and v, which eliminate the coordinate singularity at $r = 2M$. In the external region I they are related to the Schwarzschild coordinates r and t by

$$u = \sqrt{\frac{r}{2M} - 1}\; e^{r/4M} \cosh\left(\frac{t}{4M}\right),\qquad (5.9a)$$

$$v = \sqrt{\frac{r}{2M} - 1}\; e^{r/4M} \sinh\left(\frac{t}{4M}\right).\qquad (5.9b)$$

The Schwarzschild metric in terms of these new coordinates reads

$$ds^2 = \frac{32M^2}{r} e^{-r/2M}(-dv^2 + du^2) + r^2(d\theta^2 + \sin^2\theta\, d\phi^2),\qquad (5.10)$$

where $r = r(u, v)$ is determined by inverting (5.9a) and (5.9b). It is evidently regular for $r \to 2M$ and $t \to \pm\infty$, where u and v may remain finite. The

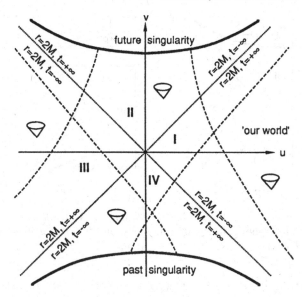

Fig. 5.2. Completed Schwarzschild solution represented in terms of Kruskal coordinates. Forward light cones now appear everywhere with a 45° opening angle around the $+v$-direction. Horizons are indicated by *dense-dotted lines*, possible orbits as *dashed lines*. Although the future horizon, say, moves in the outward direction with the speed of light from an inertial point of view, it does not increase in size. The center of the Kruskal diagram defines an 'instantaneous sphere' as a symmetry center, even though it does not specify a specific *external* time t_0

Kruskal coordinates are thus chosen in such a way that future light cones everywhere form an angle of 45° around the $+v$-direction (see Fig. 5.2). Sector I is again the external region outside the Schwarzschild radius ('our world'). One also recognizes the two distinct internal regions II and IV (connected only through the 'asymptotic sphere' at $t = \pm\infty$ that corresponds to the origin of the figure) with their two separate singularities $r = 0$. Both Schwarzschild surfaces are light-like, and thus represent one-way passages for physical orbits. Their interpretation as past and future horizons is now evident. Sector III represents the second asymptotically flat 'universe'. (It is *not* connected with the original one by a rotation in space, since u is not restricted to positive values like a radial coordinate.)

This vacuum solution of the Einstein equations is clearly T-symmetric, that is, symmetric under reflection at the hyperplane $v = 0$ (or any other hyperplane $t = $ constant). Therefore, it does not yet represent a black hole, and it would not be compatible with the Weyl tensor hypothesis. In the absence of gravitational sources, the Ricci tensor must vanish according to the Einstein equations (5.7), while a non-zero or even singular curvature tensor can then only be due to the Weyl tensor.

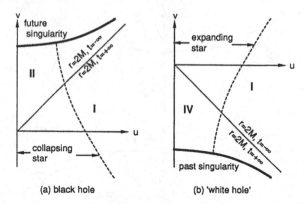

Fig. 5.3. Geometry of a Schwarzschild *black hole* (**a**) which forms by the gravitational collapse of a spherically symmetric mass, and its time-reverse (**b**) – usually called a *white hole*

A black hole is instead defined as an asymmetric spacetime structure that *arises* dynamically by the gravitational collapse of matter from a regular initial state. For example, if the in-falling geodetic sphere indicated by the dashed line passing through sectors I and II of Fig. 5.2 represents the collapsing surface of a spherically symmetric star, the vacuum solution is valid only outside it. The coordinates u and v can then be extended into the interior only with a different interpretation (see Fig. 5.3a, where $u = 0$ is chosen as the center of the collapsing star). This black hole is drastically asymmetric under time reversal, as it contains only a future horizon and a future singularity.

Because of the symmetry of the Einstein equations, a time-reversed black hole – not very appropriately called a *white hole* (Fig. 5.3b) – must also represent a solution. However, its existence in Nature would be excluded by the Weyl tensor hypothesis. If it were the precise mirror image of a black hole, the white hole could describe a star (perhaps with planets carrying time-reversed life) emerging from a past horizon. This would be inconsistent with an arrow of time that is valid everywhere in the external region. If a white hole were allowed to exist, we could receive light from its singularity, although this light would be able to carry *retarded information* about the vicinity of the singularity only if our arrow of time remained valid in this region. This seems to be required for thermodynamical consistency, but might be in conflict with such a local initial singularity (see Sect. 5.3).

Similar to past singularities, space-like singularities – so-called *naked singularities* – could also be visible to us. They, too, were assumed to be absent by Penrose. However, this 'cosmic censorship' assumption cannot generally be imposed directly as an initial condition. Rather, it has to be understood as a conjecture about the nature of singularities which may *form dynamically* during a collapse from generic initial value data which comply with the Weyl tensor hypothesis. Although counterexamples (in which naked singularities form during a gravitational collapse from appropriate initial conditions) have been explicitly constructed, they seem either to form sets of measure zero (which could be enforced by imposing exact symmetries that may be thermodynamically unstable in the presence of quantum matter fields), or to remain hidden behind black hole horizons (see Wald 1997, Brady, Moss and Myers 1998). The first possibility is similar to pathological solutions in mechanical systems that have been shown to exist as singular counterexamples to ergodic behavior. This similarity may already indicate a relationship between these aspects of general relativity and statistical thermodynamics.

The Schwarzschild–Kruskal metric may be generalized as a *Kerr–Newman metric*, which describes axially symmetric black holes with non-vanishing angular momentum J and charge Q. This solution is of fundamental importance, since its external region characterizes the final stage of *any* gravitationally collapsing object. For $t \to +\infty$ (although very soon in excellent approximation during a stellar collapse) every black hole may be completely described by the three parameters M, J and Q, up to translations and Lorentz transformations.

This result is known as the *no-hair theorem*. It means that black holes cannot maintain any external structure ('no hair'), since the collapsing star must radiate away all higher multipoles of energy and charge, while conserved quantities connected with short-range forces, such as lepton or baryon number, disappear behind the horizon. A *white hole* would therefore require coherently incoming (advanced) radiation in order to 'grow hair'. For this reason, *white holes seem to be incompatible with the radiation arrow of our world*. A general correlation between the time arrow of horizons and that of radiation has been derived in the form of a 'consistency condition' for certain de Sitter-type universes by Gott and Li (1997). Their model (though not representative for our world) is remarkable in possessing *different* arrows of time in different spacetime regions separated by an event horizon.

If the internal region of a black hole is regarded as *irrelevant* for external observers, the gravitational collapse effectively violates baryon and lepton number conservation. *Even the entropy* carried by collapsing matter would disappear from this point of view – in violation of the Second Law. Conservation laws would eventually have to be violated objectively at the future singularity if all physical properties were assumed to disappear from existence there. According to rather moderate assumptions, such a singularity must always arise behind any future horizon that comes into being (see Hawking and Ellis 1973).

Spacetime singularities would have particularly dramatic consequences in quantum theory because of the latter's kinematical nonlocality (see Sect. 4.2). Consider a global quantum state, propagating on space-like hypersurfaces ('simultaneities'), which define an arbitrary foliation of spacetime, that is, a time coordinate t. If these hypersurfaces met a singularity somewhere, not only the state of matter *on* this singularity, but also its entanglement with the rest of the Universe would be lost. While classical correlations occur only in statistical ensembles, quantum states would objectively cease to exist also on the non-singular part of the Universe unless the global state evolved 'just in time' into the factorizing form $\psi = \psi_{\text{singularity}} \psi_{\text{elsewhere}}$ whenever it approached a singularity. This would represent a very strong final condition. Therefore, several authors have concluded that quantum gravity must violate CPT invariance and unitarity, while this suggestion has led to a number of proposals for a gravity-based dynamical collapse of the wave function (Wald 1980, Penrose 1986, Károlyházy, Frenkel and Lukácz 1986, Diósi 1987, Ellis, Mohanty and Nanopoulos 1989, Percival 1997, Hawking and Ross 1997). According to these proposals, the existence of future singularities would explain the first step of (4.54).

However, this dynamical indeterminism of global quantum states would not only be inconsistent with canonical quantum gravity (see Sect. 6.2). It may also be avoided in quantum field theory on a classical spacetime if the foliation defining a time coordinate were chosen never to encounter a singularity. For example, the Schwarzschild–Kruskal metric could be foliated according to Schwarzschild time t in the external region, and according to the new time coordinate $r < 2M$ for $t > \infty$ (see Fig. 5.1). This choice, which leaves the entire black hole interior in the 'global future' of external observers *for all times*, is facilitated by the fact that this interior never enters their past, and therefore cannot be regarded as *causing* anything on them – no matter how long they wait (Zeh 2005a).

A *general* singularity-free foliation is given by *York time*, which is defined by hypersurfaces of uniform extrinsic spatial curvature scalar K (see Qadir and Wheeler 1985). A foliation that excludes singularities also appears appropriate because consequences of an elusive unified theory are expected to become relevant close to them. Many hypothetical theories have been proposed, which replace the singular big bang by other scenarios. Among them are oscillating universes or inflationary 'bubble universes' in an eternal inhomogeneous superuniverse. It is questionable, though, whether the traditional concept of time can be maintained in a situation where (quasi-)classical general relativity breaks down (see Sect. 6.2).

The conceivable salvation of *global* unitarity by excluding future singularities is quite irrelevant for local observers who remain outside the event horizon, since the reality accessible to them can be completely described by a reduced density matrix ρ_{ext} in the sense of a Zwanzig projection $\hat{P}_{\text{sub}}$ – see (4.28) – regardless of how their local reference frame is globally extrapolated to form a complete foliation. The non-unitary dynamics of these reduced density matrices has the same origin as it did for quantum mechanical subsystems of Sect. 4.3: nonlocal entanglement. Thereby, the horizon appears as a maximal boundary separating subsystems of interest. One may therefore appropriately describe the phenomenological properties of black holes (including their Hawking radiation – see below) without referring to the singularity or the precise nature of quantum gravity. This 'effective non-unitarity' of black holes reflects the usual time arrow of decoherence (see Chap. 4 and Sect. 6.2.3).

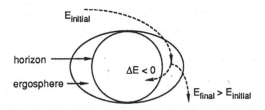

Fig. 5.4. Extraction of rotational energy from a black hole by means of the *Penrose mechanism*, using a booster in the *ergosphere* close to the horizon

From the point of view of an external observer, the information about matter collapsing under the influence of gravity becomes irreversibly irrelevant, except for the conserved quantities M, J and Q. However, the mass of a Kerr–Newman black hole is not completely lost (even if Hawking radiation is neglected). Its rotational and electromagnetic contributions can be recovered by means of a process discovered by Penrose (1969) – see Fig. 5.4. It requires boosting a rocket in the 'ergosphere', that is, in a region between the Kerr–Newman horizon, $r_+ := M + \sqrt{M^2 - Q^2 - (J/M)^2}$, and the 'static limit', $r_0(\theta) := M + \sqrt{M^2 - Q^2 - (J/M)^2 \cos^2\theta}$. In this ergosphere, the cyclic coordinate ϕ is time-like ($g_{\phi\phi} < 0$) as a consequence of extreme relativistic frame dragging. Because of the properties of this metric, ejecta from the booster which fall into the horizon may possess negative energy with respect to an asymptotic frame (even though this energy is locally positive). The mass of the black hole may thus be reduced by reducing its angular momentum. Similar arguments hold with respect to electric charge if the ejecta carry charged particles with an appropriate sign.

The efficiency of this process for extracting energy from a black hole is limited – precisely as it is for a heat engine. According to a geometro-dynamical theorem (Hawking and Ellis 1973), the area A of a future horizon (or the sum of several such horizon areas) may never decrease. For all known processes which involve black holes, this can be formulated in analogy to thermodynamics as (Christodoulou 1970)

$$dM = dM_{\text{irrev}} + \Omega\, dJ + \Phi\, dQ \,, \tag{5.11}$$

where the 'irreversible mass change' $dM_{\text{irrev}} \geq 0$ is defined by the change of total horizon area, $dM_{\text{irrev}} = (\kappa/8\pi)dA$ – in analogy to $T dS$. Here, κ is the *surface gravity*, which turns out to be constant on each horizon, while Φ is the electrostatic potential at the horizon, and Ω the angular velocity defined by the dragging of inertial frames at the horizon. The last two terms in (5.11) describe work done reversibly on the black hole by adding angular momentum or charge. All quantities are defined relative to an asymptotic rest frame, where they remain regular even when they diverge locally on the horizon. For a Schwarzschild metric, the surface gravity is $\kappa = 1/4M$. The quantities Φ and Ω are also constant on the horizon, in analogy to other thermodynamical equilibrium parameters, such as pressure and chemical potential, which appear in the expression for the work done on a thermodynamical system in the form $\mu dN - pdV$.

These similarities led to the proposal of the following *Laws of Black Hole Dynamics*, which form an analogy to the Laws of Thermodynamics (see Bekenstein 1973, Bardeen, Carter and Hawking 1973, Israel 1986):

0. The surface gravity of a black hole must approach a uniform equilibrium value $\kappa(M, Q, J)$ on a black hole horizon for $t \to \infty$.
1. The total energy of black holes and external matter, measured from asymptotically flat infinity, is conserved.
2. The total horizon area, $A := \sum_i A(M_i, Q_i, J_i)$, never decreases:

$$\frac{\mathrm{d}A}{\mathrm{d}t} \geq 0 . \tag{5.12}$$

3. It is impossible to reduce the surface gravity to zero by a finite number of physical operations.

Other versions of the Third Law of thermodynamics may not possess a direct analog in black holes because of the latters' negative heat capacity. In particular, the surface area A does not vanish with vanishing surface gravity in a similar way as the entropy does with vanishing temperature.

Bekenstein conjectured that these analogies are not just formal, but indicate genuine thermodynamical properties of black holes. He proposed not only a complete *equivalence of thermodynamical and spacetime-geometrical laws and concepts*, but even their *unification*. In particular, in order to 'legalize' the transformation of thermodynamical entropy into black hole entropy A (when dropping hot matter into a black hole), he required that instead of the two separate Second Laws, $\mathrm{d}S/\mathrm{d}t \geq 0$ and $\mathrm{d}A/\mathrm{d}t \geq 0$, there is only one *Unified Second Law*

$$\frac{\mathrm{d}(S + \alpha A)}{\mathrm{d}t} \geq 0 , \tag{5.13}$$

with an appropriate constant α (in units of $k_B c^3 / \hbar G$). Its value remains undetermined from the analogy, since the term $(\kappa/8\pi)\mathrm{d}A$, equivalent to $T\mathrm{d}S$, may equally well be written as $(\kappa/8\pi\alpha)\mathrm{d}(\alpha A)$. The *black hole temperature* $T_{\mathrm{bh}} := \kappa/8\pi\alpha$ is classically expected to vanish, since the black hole would otherwise have to emit heat radiation proportional to $A T_{\mathrm{bh}}^4$ according to Stefan and Boltzmann's law. The constant α should therefore be infinite, and so should the *black hole entropy* $S_{\mathrm{bh}} := \alpha A$.

Nonetheless, Bekenstein suggested a finite value for α (of the order of unity in Planck units). This was confirmed by means of quantum field theory by Hawking's (1975) prediction of *black hole radiation*. His calculation revealed that black holes must emit thermal radiation according to the value $\alpha = 1/4$. This process may be described by means of 'virtual particles' with negative energy tunnelling from a virtual ergosphere into the singularity (York

1983), while their entangled partners with positive energy may then propagate towards infinity. (Again, all energy values refer to an asymptotic frame of reference.) The probabilities for these processes lead precisely to a black body radiation with temperature

$$T_{\mathrm{bh}} = \frac{\kappa}{2\pi} \,,$$ (5.14)

with κ in units of $\hbar/ck_B$, and therefore to the black hole entropy[1]

$$S_{\mathrm{bh}} = \frac{A}{4} \,.$$ (5.15)

The mean wavelength of the emitted radiation is of order $\sqrt{A}$.

A black hole not coupled to any quantum fields ($\alpha = \infty$) would possess zero temperature and infinite entropy, corresponding to an ideal absorber in the sense of Sect. 2.2. This result would also be obtained for *classical black body radiation*, that is, for classical electromagnetic waves in thermal equilibrium – reflecting the historically important infrared catastrophe for classical fields (Gould 1987).

According to (5.14), a black hole of solar mass would possess a temperature of no more than $T_{\mathrm{bh}} \approx 10^{-6}$ K. In the presence of a cosmic background radiation of 2.7 K, it would therefore absorb far more energy than it emits (even in the complete absence of interstellar dust). Only black holes with mass below 3×10^{-7} solar masses could effectively *lose* mass under the present conditions of the Universe (Hawking 1976). Black holes that have formed by gravitational collapse (that is, with a mass above 1.4 solar masses) require a further expansion and cooling of the Universe by a factor of almost 10^7 or more in order to be able to disappear by radiation. 'Black-and-white holes' in equilibrium with a heat bath would not possess any horizon, but according to classical general relativity require a *spatial* singularity at $r = 0$, which corresponds to a negative singular mass – signalling the need for quantum gravity (Zurek and Page 1984).

In vacuo (at $T = 0$), a black hole would eventually completely decay into thermal radiation. The resulting entropy can be estimated to be somewhat larger than that of the black hole (Zurek 1982b). Since the future horizon and the singularity would thereby also disappear, this process seems to represent a *genuine global indeterminism* – known as the 'information loss paradox'. It is remarkable, though, that no conservation laws would be violated in a Schwarzschild foliation. The diverging time dilation close to the horizon prevents all matter from ever reaching the horizon on these simultaneities, which define a global history that covers the complete external world.

---

[1] It is important to realize that this result is quite independent of the nature of existing fields. Therefore, it cannot be used to support any *specific* theory, such as M-theory, by its explicit confirmation.

Various 'absolute' resolutions of the information loss paradox have been proposed in the context of quantum gravity (see also Sect. 6.2.3). Conventional unitary quantum theory requires that the entropy of the radiation is the consequence of a Zwanzig projection which regards entanglement between decay fragments as irrelevant. In a complete nonlocal description, photons, gravitons, neutrinos and other radiation fields would all have to be entangled to form a pure total state (Page 1980), while the latter can for all practical purposes be assumed to be a thermal mixture. However, similar quantum correlations between the constituents of incoming (advanced) radiation would be dynamically relevant for a white hole to 'grow hair'.

The description of black holes by a probabilistic 'super-scattering matrix' $ (as suggested by Hawking 1976) can thus be explained by means of decoherence in a similar manner as the apparent collapse of the wave function (see Demers and Kiefer 1996, Kiefer 2004). However, an S-matrix of any kind is *not* a realistic tool for describing black holes – just as it would not be appropriate for describing any macroscopic objects, since they can never approach an asymptotic state of perfect isolation. Unitarity would then imply the superposition of many different Everett branches (in quantum gravity including different spacetimes), while symmetries and conservation laws may be broken within individual branches. In contrast, *microscopic* 'holes' – if they exist – would not possess the classical properties 'black' or 'white', which are formally analogous to the chirality of molecules (Sect. 4.3.2), but would instead have to be described by their T-symmetric or antisymmetric superpositions.

General Literature: Bekenstein 1980, Unruh and Wald 1982.

5.2 Thermodynamics of Acceleration

While the time arrow of black holes is defined by their (quasi-)classical spacetime structure, Hawking radiation requires *quantum* fields on them. It is a consequence of quantum nonlocality, facilitated by the presence of an event horizon as a separator between different 'subsystems'. However, the relevance (or even existence) of this horizon depends on the worldlines of observers or detectors, which define comoving local frames of reference. A black hole horizon is relevant for observers in its flat asymptotic spacetime, or for those staying at a fixed distance, while it would not exist for observers freely falling in. Would their detectors then register any Hawking radiation?

Homogeneous gravitational fields are known to be equivalent to uniformly accelerated frames of reference. They do not require any spacetime curvature, but can be transformed away by means of accelerated (curved) spacetime coordinates. A massive plane, for example, is equivalent to a discontinuity of inertial frames, separating the half-spaces on both sides of the plane by a uniform relative acceleration in the direction orthogonal to the plane. Must an accelerated detector in vacuo therefore be expected to register thermal

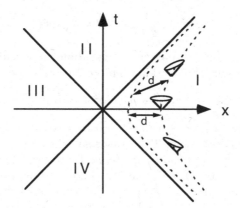

Fig. 5.5. Horizons in Minkowski spacetime are defined for uniformly accelerated local observers (*dashed* world lines) by the asymptotes of their hyperbolic world lines, which are described by the equation $\rho := (x^2 - t^2)/4 =$ constant. Proper acceleration depends on the specific world line. Distances d between two parallel observers remain constant in their comoving rest frames, thus defining global rigid frames in regions I and III

radiation 'equivalent' to (5.14)? For a uniformly accelerated observer in flat Minkowski spacetime, there would indeed be a past and a future horizon, represented by the asymptotes of his hyperbolic relativistic world line (see Fig. 5.5). He shares these horizons with a whole family of 'parallelly accelerated' observers (who require different accelerations in order to remain on parallel hyperbolae – *equivalent* to two observers at different fixed distances from a black hole). These observers also share their comoving rest frames, and thus define an accelerated global rigid frame in keeping fixed distances d in spite (or rather because) of their different accelerations. The same kinematical situation had to be discussed for uniformly accelerated charges and detectors of classical electromagnetic waves in Sect. 2.3.

The two-dimensional Minkowski diagram of Fig. 5.5 appears similar to the Kruskal–Szekeres diagram (Fig. 5.2), although it is singularity-free, as each point in Fig. 5.5 represents a flat $\mathbb{R}^2$ (with coordinates y, z) rather than a 2-sphere. Therefore, points in regions I and III are now related by a π-rotation around the t-axis. If the acceleration had *begun* at a certain finite time ($t = 0$, say), no past horizon would exist (in analogy to a black hole – see Fig. 5.3a). The world lines of this family of local observers can be used to define a new spatial coordinate $\rho(x,t)$ that is constant for each of them, and may be conveniently scaled by $\rho(x,0) = x^2/4$. Together with a new time coordinate $\phi(x,t)$ that is related to proper times τ along the world lines according to $d\tau = \sqrt{\rho}\, d\phi$, and the coordinates y and z, it defines the *Rindler coordinates* of flat spacetime. In region I of Fig. 5.5, they are related to the Minkowski coordinates by

$$x = 2\sqrt{\rho}\cosh\frac{\phi}{2} \quad \text{and} \quad t = 2\sqrt{\rho}\sinh\frac{\phi}{2} \,. \tag{5.16}$$

The proper accelerations $a(\rho)$ along $\rho = $ constant are given by $a = (2\sqrt{\rho})^{-1}$, while the resulting non-Minkowskian representation of the Lorentz metric,

$$ds^2 = -\rho\,d\phi^2 + \rho^{-1}\,d\rho^2 + dy^2 + dz^2 \,, \tag{5.17}$$

describes a coordinate singularity at $\rho = 0$ that is analogous to $r - 2M = 0$ for the Schwarzschild solution. The Minkowski coordinates can therefore be compared with the Kruskal coordinates u and v of Fig. 5.2, while the Rindler coordinates are analogous to the Schwarzschild coordinates.

The Rindler coordinates are also useful for describing the uniformly accelerated point charge of Sect. 2.3 and its relation to a co-accelerated detector. The radiation propagating along the forward light cone of an event on the accelerated world line of the charge must somewhere hit the latter's future horizon (see Fig. 5.5), and asymptotically completely enter region II. However, from the point of view of a co-accelerated (uniformly Lorentz-rotated) observer with the same comoving simultaneities $\phi = $ constant, which all intersect the horizon at the origin, the radiation would *never* reach the horizon at $\phi = \infty$.

This explains why the accelerated charge radiates from the point of view of an inertial observer, but not for a co-accelerated one (Boulware 1980). While Dirac's *invariant* radiation reaction (2.25) vanishes for uniform acceleration, the definition of radiation is based on the distinction between near fields and far fields by their dependence on different powers of distance according to (2.14), and therefore depends on the acceleration of the reference frames. Even though global inertial frames are *absolutely* defined in special relativity, it is the *relative* acceleration between source and detector that is relevant for the resulting effects. For a *uniformly* accelerated charge in region I, time reversal symmetry has the consequence that its total retarded field is identical with its advanced field in this whole sector (except on the horizons), while elsewhere one has either just the retarded outgoing fields in region II, or just the advanced incoming fields in region IV (or a superposition of these two cases) – depending on the boundary conditions.

Unruh (1976) was able to show that an accelerated detector in the inertial vacuum of a quantum field must register an isotropic thermal radiation of *all* existing fields, corresponding to the temperature

$$T_U := \frac{a}{2\pi} = \frac{a\hbar}{2\pi c k_B} \,. \tag{5.18}$$

This is precisely what had to be expected from (5.14) according to the principle of equivalence, and from the analogy with Mould's (1964) result for classical radiation. For a generalization of (5.18) to other trajectories see Louko and Satz (2006). However, the response of a detector appears as an objective fact. It cannot just be a matter of spacetime perspective or definition (such as the distinction between near field and far field in different coordinate systems): wave functions live on configuration space.

The result (5.18) can be understood when representing the inertial *Minkowski vacuum* $|0_M\rangle$ in terms of 'Rindler modes', that is, wave modes which factorize in the Rindler coordinates (with frequencies Ω with respect to the time coordinate ϕ). If Minkowski plane wave modes $e^{i(kx-\omega t)}$ are expanded in terms of such Rindler modes, this leads to a *Bogoljubow transformation* for the corresponding 'particle' creation operators:

$$a_k^+ \longrightarrow b_{\Omega s}^+ := \sum_k (\alpha_{\Omega s,k} a_k^+ + \beta_{\Omega s,k} a_k) .$$

Here, the index $s = I$ or III specifies two Rindler modes (both with time dependence $e^{-i\Omega\phi}$) which vanish in the regions III or I of Fig. 5.5, respectively. On flat simultaneities through the origin ($\phi = $ const.), they are complete on the corresponding half-spaces with $x > 0$ or $x < 0$, respectively. These Bogoljubow transformations combine creation and annihilation operators, since the non-linear coordinate transformations (5.16) do not preserve the sign of frequencies. These signs distinguish particle and antiparticle modes in the usual interpretation, such that the two terms of the Fourier representation of field operators, $\Phi(r,t) \propto \int \{ \exp[i(kx + \omega t)]a_k + \exp[i(kx - \omega t)]a_k^+ \}dk$, are not separately transformed. (Recall that 'particle creation' operators are just raising operators for harmonic oscillator quantum numbers characterizing quantum states of field modes.)

In terms of the Rindler modes, the Minkowski vacuum becomes an *entangled* state in the form of a BCS ground state of superconductivity (Bardeen, Cooper and Schrieffer 1957):

$$|0_M\rangle = \prod_\Omega \left(\sqrt{1 - e^{-4\pi\Omega}} \sum_n e^{-2\pi\Omega n} |n\rangle_{\Omega,I} |n\rangle_{\Omega,III} \right) , \qquad (5.19)$$

where $|n\rangle_{\Omega,s} = (n!)^{-1/2}(b_{\Omega s}^+)^n |0_R\rangle$ are the Rindler particle occupation number eigenstates (see also Gerlach 1988). The *Rindler vacuum* $|0_R\rangle$, defined by $b_{\Omega s}|0_R\rangle = 0$ for all Ω and s, is therefore different from the Minkowski vacuum. It must also be a *pure* state in the Minkowski representation, while its reduced density matrix on the half spaces $x > 0$ or $x < 0$ describes a thermal mixture. This demonstrates that the concepts of quantum 'particles' and their vacua are not invariant under non-Lorentzian transformations. While the *actual quantum state* may be regarded as absolutely defined ('real'), its interpretation in terms of 'particles' depends on the local choice of simultaneities – conveniently identified with the comoving rest frames of a detector. For example, the Rindler basis characterizes detectors accelerated relative to inertial frames, while a specific 'vacuum' would represent an actual (physically meaningful) state. This distinction between physical states and their various representations is obscured in the Heisenberg picture.

Equation (5.19) is the *Schmidt canonical representation* (4.27) of nonlocal quantum correlations between the two sectors I and III of Fig. 5.5 (which together are spatially complete for hyperplanes intersecting the origin at $x = t = 0$). It illustrates the kinematical nonlocality of a relativistic Minkowski vacuum. The diagonal elements represent a canonical distribution with dimensionless formal temperature $1/4\pi$ – compatible with the dimensionless time coordinate ϕ. Since proper times along the world lines $\rho, y, z =$ constant are given by $d\tau = \sqrt{\rho}\, d\phi = (2a)^{-1}\, d\phi$, energies are given by $2an\Omega$. The (ρ-dependent) temperature is therefore $T = a/2\pi$ – in accordance with (5.18). Disregarding quantum correlations with the other half-space thus leads to the *apparent ensemble* of states representing a heat bath. As one needs measurement times Δt larger than $(a\Omega)^{-1}$ to measure a frequency Ω, the acceleration has to remain approximately uniform for more than this interval of time in order to mimic the presence of an event horizon for this mode.

While the result (5.18) might have been expected from the principle of equivalence, it is more general than (5.14), since it is independent of gravity (spacetime curvature). In general, the equivalence principle holds only locally. Its exceptional global applicability is a consequence of the specific field of uniform accelerations depicted in Fig. 5.5 (see also Sect. 2.3). Therefore, Unruh radiation cannot in general be *globally* equivalent to Hawking radiation. While the whole future light cone of an event on the world line of a uniformly accelerated object must asymptotically intersect the latter's horizon for an inertial observer (as discussed above), only *part* of the future light cone of an event in the external region of a black hole will ever enter its internal region. For an observer approaching a black hole, the horizon will eventually cover his whole celestial sphere because of the bending of light rays. (He would have to speed towards the remaining 'hole in the sky' in order not to be swallowed.) Such spacetime-geometric aspects of boundary conditions also determine the specific 'actual vacuum' (Unruh 1976). Only in the immediate neighborhood of the horizon can the freely falling observer be completely equivalent to the inertial one in flat spacetime, and thus precisely experience a vacuum. While the Unruh radiation is isotropic and T-symmetric, the Hawking radiation observed by a non-inertial detector at a fixed distance from a black hole specifies a direction in space as well as in time by its non-vanishing energy flux coming from the black hole.

A real (and in principle observable) accelerated QED vacuum could be produced by a uniformly accelerated ideal mirror (Davies and Fulling 1977). A mirror at rest, representing a plane boundary condition to the field, leads to the removal of an infinite number of field modes (those not matching the boundary condition). This in turn leads to an infinite energy renormalization (defining a 'dressed mirror') by subtracting their zero point energies. This dressing would not be additive for two or more parallel mirrors at fixed distances, while the adiabatic variation of their distances defines the finite and observable *Casimir effect* (a force between them). An *accelerated* mirror, acting as an accelerated boundary, produces a quantum field state that would be experienced as a vacuum by a co-accelerated detector, but as a thermal bath by an inertial one. A uniformly accelerated mirror would completely determine this QED state on the concave side of its spacetime hyperbola in Fig. 5.5, while the convex side offers the freedom of additional boundary conditions in regions II or IV (similar to the classical field of a uniformly accelerated charge). According to the equivalence principle, an 'ideal graviton mirror' would even redefine (completely 'drag') inertial frames.

All these thermodynamic consequences of acceleration or curvature are too small to be confirmed with presently available techniques. However, they were drawn by combining two well established theories (general relativity and quantum field theory), and they appear necessary for consistency (see Unruh and Wald 1982). So they can hardly be regarded as merely hypothetical.

General Literature: Birrell and Davies 1983.

5.3 Expansion of the Universe

Since Hubble's discovery of 1923, we have known that the Universe is expanding. This is often regarded as a confirmation of general relativity, since it can be described by Friedmann's solutions of the Einstein equations of 1922. However, a very similar dynamical universe could have been derived in Newton's theory, although this nonrelativistic model would have to specify an inertial center. Evidently, applying the laws of mechanics and gravity to the whole Universe met even stronger reservations than applying them to the celestial objects a few hundred years earlier, when Kepler and his contemporaries were surprised to discover that planets can 'fly like the birds' rather than being guided by the crystal spheres.

Since a static universe would not be stable under gravity, Einstein quite artificially introduced his 'cosmological constant' in order to make his theory compatible with what he believed to be empirically correct. A similar novel kind of repulsive global force would have been required in Newton's theory for this purpose. In an *open* Newtonian universe these consequences might at most be obscured, but not avoided (Bondi 1961). Without such a repulsive force, Newton's theory, too, would have required the Universe to expand or to contract (depending on the initial conditions), and this would have led to a big bang or a big crunch, respectively, or both. However, in contrast to general relativity, a singularity could then be avoided by appropriate repulsive forces that become relevant at very high densities.

In Einstein's theory, a homogeneous and isotropic universe is described by the Friedmann–Robertson–Walker (FRW) metric,

$$ds^2 = -dt^2 + a(t)^2 \left\{ d\chi^2 + \Sigma^2(\chi) \left[d\theta^2 + \sin^2\theta \, d\phi^2 \right] \right\}, \tag{5.20}$$

with $\Sigma(\chi) = \sin\chi$, $\sinh\chi$, or χ, depending on the sign of the spatial curvature, $k = +1$, -1 or 0, respectively. The Friedmann time coordinate t in (5.20) describes the proper time for objects which are at rest in these coordinates ('comoving clocks'). This metric may remain valid close to the big bang (for $a = 0$) in accordance with the Weyl tensor hypothesis. It can be generalized by means of a multipole expansion on the Friedmann sphere (see Halliwell and Hawking 1985, and Sect. 5.4). This general-relativistic form has the advantages of not requiring a special 'center at rest', and of allowing a finite universe without a boundary (for positive curvature).

The exact FRW metric (5.20) depends only on the expansion parameter $a(t)$. The latter's dynamics, derived from the Einstein equations (5.7) with an additional cosmological constant, assumes the form of an 'energy integral' with a *fixed vanishing* value of the energy:

$$\frac{1}{2}\left(\frac{1}{a}\frac{da}{dt}\right)^2 = \frac{1}{2}\left(\frac{d\alpha}{dt}\right)^2 = -V(\alpha)\,. \qquad (5.21)$$

The logarithm of spatial extension, $\alpha = \ln a$, which formally sends the big bang to minus infinity, will prove convenient on several occasions. The Friedmann potential $V(\alpha)$ is given by the energy density of matter $\rho(a)$, the cosmological constant Λ, and the spatial curvature k/a^2, in the form

$$V(\alpha) = -\frac{4\pi\rho(e^\alpha)}{3} - \frac{\Lambda}{3} + ke^{-2\alpha}\,. \qquad (5.22)$$

One would have obtained essentially the same equation (without curvature term and cosmological constant, but with variable energy) from Newton's dynamics for the radius of a gravitating homogeneous sphere of matter.

The energy density ρ may depend on a in various ways. In the matter-dominated epoch it is proportional to the inverse density, a^{-3}. During the radiation era – less than 10^{-4} of the present age of the universe – it decreased according to a^{-4}, since all wavelengths expand with a. Much earlier (for extremely high matter density), quite novel phenomena must be expected to have affected the relativistic equation of state, here described by $\rho(a)$. According to some theories, for example, the vacuum state of matter passed through one or several phase transitions (see Sect. 6.1). Similar to a condensation process, this situation may be characterized by a constant function of state, $\rho(a) = \rho_0$. The matter term in the potential V would then *simulate* a cosmological constant – albeit only for a limited time (see Fig. 5.6). In the 'Planck era', that is, for values of a of order unity, quantum gravity must become essential (see Sect. 6.2).

Different eras, described by such analytic equations of state $\rho(a)$, possess different solutions $a(t)$. For example, a dominating (fundamental or simulated) cosmological constant would lead to a 'de Sitter era' with $a(t) = ce^{\pm Ht}$ and a 'Hubble constant' $H = \dot{a}/a = \dot{\alpha}$. For a matter- or radiation-dominated universe, one has $a(t) = c't^{2/3}$ or $a(t) = c''t^{1/2}$, respectively, while for low matter densities the curvature term may dominate. Recent observations indicate that our Universe is approximately flat (negligible curvature term), while an effec-

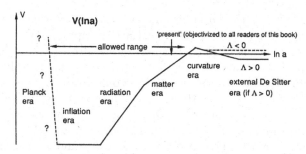

Fig. 5.6. Schematic behavior of the 'potential energy' for the dynamics of $\ln a$ in the case of positive spatial curvature. Since only regions with positive kinetic energy $E - V = -V > 0$ are allowed, turning points of the cosmic expansion would arise at values of $V = 0$. An upper turning point would lead to recontraction, while a lower turning point describes a 'bouncing' universe (without big bang or big crunch singularities)

tive positive cosmological constant of unknown origin ('dark energy') already contributes two thirds of the potential $V(a)$.

5.3.1 Instability of Homogeneity

While the Friedmann model is an exact solution of the Einstein equations, and apparently a reasonable approximation to the very large scale behavior of the real Universe, it is not stable against density fluctuations (as discussed in the introduction to this chapter and in Sect. 5.1). This local instability cannot be compensated by a *global* force, such as a cosmological constant. It is in fact successfully used to explain the formation of stars, galaxies, galaxy clusters, possibly larger structures, and eventually black holes in the present Universe. Thereby, the *assumed* initial symmetries of the Friedmann universe must be dynamically broken. In classical physics, density fluctuations would be microscopically determined (Sect. 3.4). In quantum theory they may also result from an indeterministic (genuine or apparent) collapse of the wave function, induced by decoherence (see Calzetta and Hu 1995, Kiefer, Polarski and Starobinsky 1998, and Sect. 6.1). A similar quantum effect is known to limit the retardation of symmetry-breaking phase transitions (their hysteresis). The onset of these primordial structures of the Universe is now believed to be observed in the cosmic background radiation.

The arrow of time characterizing these irreversible processes is thus again based on an improbable (but 'simple') cosmic initial condition: homogeneity. When Boltzmann (1896) discussed the origin of the Second Law in the context of an infinite and eternal universe, he had to conclude that we, here and now, are living in the aftermath of a gigantic cosmic fluctuation. Its maximum (that is, a state of very low entropy) must have occurred in the distant past in order to explain the existence of fossils and other documents in terms of causal history and evolution (see Sect. 3.5).

How improbable is the novel initial condition of homogeneity that Boltzmann did not even recognize as an essential assumption? We may calculate its probability by means of Einstein's relation (3.56) if we know the entropy of the most probable state. The entropy of a non-degenerate homogeneous physical state in local equilibrium is proportional to the number of particles, N. All other parameters enter this expression only logarithmically – as exemplified for the ideal gas in (3.14). In the present Universe, the number of photons contained in the 2.7 K background radiation exceeds that of massive particles by a factor 10^8. The entropy of a finite 'standard universe' of 10^{80} baryons (now often regarded as no more than a 'bubble' in a much larger or infinite universe) would therefore possess an entropy of order 10^{88} plus a small but important contribution resulting from gravitating objects. Most of this entropy must therefore have been produced in the early Universe by the creation of photons and other particles, which are strongly entangled in a chaotic way.

However, the present entropy is far from its maximum that would be achieved by the production of black holes. In Planck units, the horizon area of a neutral and spherical black hole of mass M is given by $A = 4\pi(2M)^2$. Its entropy according to (5.15) thus grows with the *square* of its mass,

$$S_{\text{bh}} = 4\pi M^2 \ . \tag{5.23}$$

Merging black holes will therefore produce an enormous amount of entropy. If the standard universe of 10^{80} baryons consisted of 10^{23} solar mass black holes (since $M_{\text{sun}} \approx 10^{57} m_{\text{baryon}}$), it would already possess a total entropy of order 10^{100}, that is, 10^{12} times its present value. If most of the matter eventually formed a single black hole, this value would increase by another factor of 10^{23}. The probability for the present, almost homogeneous universe is therefore a mere

$$p_{\text{hom}} \approx \frac{\exp(10^{88})}{\exp(10^{123})} = \exp(10^{88} - 10^{123}) \approx \exp(-10^{123}) \tag{5.24}$$

(Penrose 1981), indistinguishable in this approximation from the much smaller probability at the big bang. Gravitational contraction thus offers an enormous further entropy capacity to assist the formation of structure and complexity.

This improbable initial condition of homogeneity as an origin of thermodynamical time asymmetry is different from attempts (see Gold 1962) to derive this arrow from a homogeneous expansion of the Universe in a causal manner (see Price 1996 and Schulman 1997 for critical discussions). While it is true that non-adiabatic expansion of an equilibrium system may lead to a retarded non-equilibrium, this would equally apply to non-adiabatic *contraction* in our causal world. The growing space (and thus phase space, representing increasing entropy capacity) cannot form the master arrow of time, since it is insufficient to explain causality (the absence of any advanced correlations). Non-adiabatic *compression* of a vessel would lead to retarded pressure waves emitted from the walls, but not to a reversal of the thermodynamical arrow. The entropy capacity of gravitational contraction is far more important than homogeneous expansion, but probably not very relevant for the very early stages of the Universe.

There are other examples of *using* causality in thermodynamical arguments rather than *deriving* it in this cosmic scenario. For example, Gal-Or (1974) discussed retarded equilibration due to the slow nuclear reactions in stars. Even though nuclear fusion controls the time scale and energy production during most stages of stellar contraction, it presumes a strong initial non-equilibrium.

5.3.2 Inflation and Causal Regions

The finite age of an expanding universe that starts from an initial singularity (a big bang) leads to the consequence that the backward light cones of two events may not overlap. These events would then not be causally connected. A sphere formed by the light front originating in a point-like event at the big bang, where $a(0) = 0$, is therefore called a *causality horizon*. Its radius $s(t)$ at Friedmann time t is given by

$$s(t) = \int_0^t \frac{a(t)}{a(t')} \mathrm{d}t' \, . \tag{5.25}$$

In a matter- or radiation-dominated universe, this integral would converge for $t' \to 0$, and thus define a finite horizon size. Only *parts* of the Universe may then be causally connected – excluding even readily observable distant pairs of objects that strongly indicate a simultaneous origin.

In particular, the homogeneity of the universe on the large scale would thereby remain causally unexplained. This *horizon problem* was the major motivation for postulating a phase transition of the vacuum or another mechanism of quantum fields that would lead to a transient cosmological constant, and thus to an early de Sitter era. In an exponentially expanding universe, the big bang singularity could in principle be shifted arbitrarily far into the past – depending on the duration of this era. However, in an extremely short time span (of the order of 10^{-33} s), the universe, and with it all causality horizons, would have been *inflated* by a huge factor that was sufficient for the sources of the whole now observable cosmic background radiation to be causally connected (Linde 1979). On the other hand, since causality horizons started with zero radius, this would explain the initial absence of nonlocal correlations and entanglement, provided they were assumed to *require* a causal origin.

Measurements of the cosmic background radiation indicate that an inflation era did in fact occur. Since the corresponding repulsive force counteracts gravity, it has also been conjectured to have driven the universe into a state of homogeneity in a causal manner. This *cosmological no-hair conjecture* is supported by a theorem of Hawking and Moss (1982). However, this theorem remains insufficient for the required purpose, since the global effect of a cosmological constant cannot generally force *local* gravitating systems, in particular black holes, to expand into a state of homogeneity. Proofs of the cosmic no-hair theorem had therefore to exclude positive spatial curvature. (Expanding white holes would require acausally incoming advanced radiation, as explained in Sect. 5.1.)

Since a cosmological constant that was simulated by a phase transition of the vacuum would depend on the local density, it may at least overcompensate the effect of gravity until strong inhomogeneities begin to form. This may *partly* explain the homogeneity of the observed part of our universe. It can be described by saying that the Weyl tensor 'cooled down' as a consequence of this spatial expansion – similar to the later red-shifting of the primordial electromagnetic radiation. While these direct implications of the expansion of the universe define reversible phenomena, equilibration during the radiation era or during the phase transition would be irreversible in the statistico-thermodynamical sense (based on microscopic causality).

This explanation of homogeneity is incomplete as it has to presume the absence of *strong* initial inhomogeneities (abundant initial black holes, in particular). In order to work in a deterministic theory, it would furthermore require the state that precedes inflation to be even less probable than the homogeneous state after inflation.

Similar inflationary scenarios have been discussed in various hypothetical models of quantum cosmology (see Caroll and Chen 2004, and Chap. 6).

5.3.3 Big Crunch and a Reversal of the Arrow

These questions may also be discussed by means of a conceivable recontracting universe. A consistent analysis of the arrow of time for this case is helpful regardless of what will happen to our own Universe. Would the thermodynamical arrow have to reverse direction when this universe starts recontracting towards the big crunch after having reached maximum extension? The answer would have to be 'yes' if the cosmic expansion represents the master arrow, but it is often claimed to be 'no' on the basis of causal arguments if they are continued into this region. For example, some authors argued that the background radiation would reversibly heat up during contraction (blueshifting), while the temperature gradient between interstellar space and the fixed stars would first have to be inverted in order to reverse stellar evolution long after the universe had reached its maximum extension. However, this argument presupposes the overall validity of the 'retarded causality' in question, that is, the absence of future-relevant correlations in the contraction phase. It would be justified if the relevant initial condition held at only one 'end' of this otherwise symmetric cosmic history. The absence or negligibility of any anti-causal events in our present epoch seems to indicate either that our Universe is thermodynamically asymmetric in time, or that it is still 'improbably young' in comparison to its total duration.

Paul Davies (1984) argued in a similar causal manner that there can be no reversed inflation leading to a homogeneous big crunch, since correlations which would be required for an inverse phase transition have to be excluded for being extremely improbable. Instead of a homogeneous big crunch one would either obtain locally re-expanding 'de Sitter bubbles' forming an inhomogeneous 'bounce', or inhomogeneous singularities at variance with a reversed Weyl tensor condition, or both. This probability argument fails, however, if the required correlations are *caused in the backward direction of time* by a final condition that was thermodynamically a mirror image in time of the initial one (see also Sect. 6.2.3). Similarly, if the big bang was replaced by a non-singular *homogeneous bounce* by means of some kind of 'Planck potential' (Fig. 5.6), entropy must have decreased prior to the bounce. In particular, decoherence would have to be replaced by recoherence in all contraction eras. In this case, an observer complying with the Second Law would always experience an expanding universe; the sign of the dynamical time parameter used in this description is merely formal (see Sect. 5.4).

On the other hand, a low entropy big bang *and* an equivalent big crunch may lead to severe consistency problems, since the general boundary value problem (Sect. 2.1) allows only one complete (initial or final) condition. Although the requirement of low entropy is not a complete boundary condition, statistically independent two-time conditions would lead to the square of the already very small probability of (5.24), that is,

$$p_{\text{two-time}} = p_{\text{hom}}^2 \approx \left[\exp(-10^{123}) \right]^2 \approx \exp(-10^{123.301}) \ . \tag{5.26}$$

The RHS appears as a small correction to (5.24) only because of this double-exponential form, although an element of phase space corresponding to (5.26) could now easily be much smaller than a Planck cell (see Zeh 2005b). A two-time boundary condition of homogeneity may thus be inconsistent with 'ergodic' quantum cosmology (that would have to include the repeated formation and decay of black holes, which contribute most of phase space).

The consistency of general two-time boundary conditions has been investigated for simple deterministic systems (see Cocke 1967 and Schulman 1997). Davies and Twamley (1993) discussed the more realistic situation of classical electromagnetic radiation in an expanding and recollapsing universe. According to their estimates, our Universe will remain essentially transparent all the way between the two opposite radiation eras (in spite of the reversible red- and blue-shifting over many orders of magnitude in between) − in contrast to ergodic assumptions used in (5.26). Following a suggestion by Gell-Mann and Hartle, they concluded that light emitted *causally* by all stars before the 'turning of the tide' propagates freely until it reaches the time-reversed radiation era − thus giving rise to an asymmetric history of this universe.

David Craig (1996) argued on this basis, but by *assuming* a thermodynamically time-symmetric universe, that the night sky at optical frequencies should contain an almost homogeneous component that represents the advanced radiation from stars existing during the contraction era. It should be observable as a non-Planckian high frequency tail in the isotropic background radiation with a total intensity at least equalling that of the light now observed from all stars and galaxies in our past – but probably much higher because of the advanced light corresponding to that which will have to be produced until the turning point is reached. However, since classical radiation would preserve all information about its origin, it is inconsistent with a time-reversed absorber (the opposite radiation era), that allows only its thermal radiation in *its* causal future (Sect. 2.2). Craig also concluded that the intensity of the thermal part of the background radiation would be doubled because of the two radiation eras, but this does not seem to be required, since the 'two' *thermal* components may be identical. (Retarded and advanced fields do not add – see Sect. 2.1 – but they must be consistent with one another.) Only in the non-thermal frequency range can retarded and advanced radiation be conceptually distinguished and thus carry information about their origin.

These conclusions have to be modified in an essential way when the quantum aspect of electromagnetic radiation is taken into account. The information content of radiation consisting of photons is limited, as first emphasized by Brillouin (1962). This consequence had also turned out to be important for Borel's argument of Sect. 3.1.2 – see footnote 4 of Chap. 3. Each photon, even if emitted into intergalactic space as a spherical wave, disappears from the whole quasi-classical universe as soon as it is absorbed *somewhere*. A reversal of this process would again require recoherence, that is, the superposition of many Everett branches. This argument requires consistent quantum cosmology (Chap. 6), where initial or final conditions can only affect the total, unitarily evolving Everett wave function. If the Schrödinger dynamics was instead modified by means of a collapse of the wave function (as implicitly assumed also for Gell-Mann and Hartle's 'histories'[2]), the corresponding new

---

[2] Gell-Mann and Hartle (1994) discussed quantum mechanical 'histories', which are defined in terms of time-ordered series of projections in Hilbert space. These *individual* histories are thus equivalent to successions of stochastic collapse events (global quantum jumps) – even though a collapse is not explicitly used. The authors nonetheless discussed the possibility of a thermodynamically time reversal-symmetric cosmic history by presuming a final condition that is similar to the initial one. This proposal is based on the equivalence of the upper and lower diagrams of Fig. 4.4, but neglects the asymmetric structure (4.56) of a collapse, which would have to include all retarded entanglement with 'information gaining systems'. Therefore, it leads to insurmountable problems as soon as one attempts to justify the probabilistic interpretation ('consistent histories') by an in practice irreversible decoherence process (see Fig. 4.5). Time reversal symmetry could be restored in the contraction era only by means of a complete process of recoherence. This would not only have to include those Everett components that have been disregarded by the Hilbert space projections which lead to individual measurement outcomes, and in this way define quasi-classical 'histories' as a *partial* quantum reality. It should also require components that have to be regarded as being retro-caused in the future.

dynamical law would have to be reversed, too, in order to save a thermodynamically time-symmetric (but now indeterministic) universe.

This problem of consistent cosmic two-time boundary conditions will assume a conceptually quite novel form in the context of quantum gravity, where
any fundamental concept of time disappears from the description of a closed
universe (Sect. 6.2).

5.4 Geometrodynamics and Intrinsic Time

In general relativity, the 'block universe picture' is traditionally preferred to a
dynamical description, as its unified spacetime concept is then manifest. So it
took almost half a century before its dynamical content was sufficiently understood, in particular by means of its Hamiltonian form, invented by Arnowitt,
Deser and Misner (1962). This approach, which is essential for a quantization of the theory, has not always been welcomed, as it seems to destroy the
beautiful relativistic spacetime concept by reintroducing a 3+1 (space and
time) representation. However, only in this *form* can the dynamical content
of general relativity be fully appreciated (see Chap. 21 of Misner, Thorne and
Wheeler 1973). A similarly symmetry-violating form in spite of Lorentz invariance is known for the electromagnetic field when described in the Coulomb
gauge by the vector potential A as the dynamical field configuration on a
space-like hypersurface of Minkowski spacetime.

This dynamical reformulation requires the separation of unphysical gauge
degrees of freedom (which in general relativity simply represent the choice of
coordinates), and the skillful handling of boundary terms. The result of this
technically demanding procedure turns out to have a simple interpretation. It
describes the *dynamics of the spatial geometry* ('three-geometry') $^{(3)}G(t)$, that
is, a propagation of the intrinsic curvature on space-like hypersurfaces with
respect to a time coordinate t that labels a foliation of the spacetime arising
dynamically in this way. This foliation has to be *chosen* simultaneously with
the construction of the solution. The extrinsic curvature, which describes the
embedding of the three-geometries into spacetime, is represented by the corresponding canonical momenta. The configuration space of three-geometries
$^{(3)}G$ has been dubbed *superspace* by Wheeler, since the form of its kinetic energy defines a metric. Trajectories in this superspace define four-dimensional
spacetime geometries $^{(4)}G$.

This 3+1 description may appear ugly not only as it hides Einstein's beautiful spacetime concept, but also since the foliation of a given $^{(4)}G$ by means
of space-like hypersurfaces, on which $^{(3)}G(t)$ is defined, is quite arbitrary.
Many trajectories $^{(3)}G(t)$ therefore represent the same spacetime $^{(4)}G$, which
is absolutely defined. It is only in special situations – such as for the FRW
metric (5.20) – that there may be a 'preferred choice' of coordinates, which
then reflect their exceptional symmetry. The time coordinate t, characterizing
a foliation, is just one of the four arbitrary (physically meaningless) spacetime

coordinates. As a parameter labelling trajectories it could just as well be eliminated and replaced by one of the dynamical variables (a global 'clock' – see Chap. 1), such as the size (or scale) of an expanding universe. The abstract four-geometry defines all spacetime distances – including *all* proper times of real or imagined local clocks. Classically, spacetime may always be assumed to be filled with a 'dust of test clocks' of negligible mass (see Brown and Kuchař 1995). However, such clocks are not required to *define* proper times; in general relativity, time as a property of the metric is itself a dynamical variable (see below), while proper times assume the role of Newton's time as controllers of motion for all material clocks.

Einstein's equations (5.7) possess a similar hyperbolic structure as the wave equation (2.1). They may therefore be expected to determine the metric $g_{\mu\nu}(x, y, z, t)$ by means of two boundary conditions for $g_{\mu\nu}$ – at t_0 and t_1, say. (For $t_1 \to t_0$ this would correspond to $g_{\mu\nu}$ and its 'velocity' at t_0. This pair of variables would in general also define the extrinsic curvature.) Since the time coordinate is physically meaningless, its value on the boundaries is irrelevant: two metric functions on three-space, $g_{\mu\nu}^{(0)}(x, y, z)$ and $g_{\mu\nu}^{(1)}(x, y, z)$, without mentioning time coordinates, suffice to determine a solution and hence physical time. Not even their order is essential, since there is no *absolute* direction of light cones. Similarly, the t-derivative of $g_{\mu\nu}$, resulting in the limit $t_1 \to t_0$, is required only up to a scalar factor (that would specify a meaningless initial 'speed of three-geometry' in superspace).

If one also eliminates all *spatial* coordinates from the metric $g_{\mu\nu}(x, y, z)$, it describes precisely the coordinate-independent three-geometry $^{(3)}G$. One may therefore expect the coordinate-independent content of the Einstein equations to determine the complete four-dimensional spacetime geometry in-between (and possibly beyond) two spatial geometries $^{(3)}G^{(0)}$ and $^{(3)}G^{(1)}$. However, the existence and uniqueness of a solution for this boundary value problem has not yet been generally proved (Bartnik and Fodor 1993, Giulini 1998).

The procedure is made transparent by writing the metric with respect to a chosen foliation as

$$\begin{pmatrix} g_{00} & g_{0l} \\ g_{k0} & g_{kl} \end{pmatrix} = \begin{pmatrix} N^i N_i - N^2 & N_l \\ N_k & g_{kl} \end{pmatrix}. \tag{5.27}$$

The submatrix $g_{kl}(x, y, z, t)$ (with $k, l = 1, 2, 3$) for $t = $ constant is now the spatial metric on a hypersurface, while the *lapse function* $N(x, y, z, t)$ and the three *shift functions* $N_i(x, y, z, t)$ define arbitrary increments of time and space coordinates, respectively, for an orthogonal transition to an infinitesimally close space-like hypersurface. These four 'gauge functions' have to be *chosen* for convenience when solving an initial value problem.

The six functions forming the remaining symmetric matrix $g_{kl}(x, y, z, t)$ still contain three gauge functions representing the spatial coordinates. Their initial choice is specified by the initial matrix $g_{kl}^{(0)}(x, y, z)$, while the free shift functions determine their change with time. The three remaining, geometrically meaningful functions may be physically understood as representing the two polarization components of gravitational waves and the 'many-fingered' (local) physical time that describes the increase of all proper times along world lines connecting two infinitesimally close space-like hypersurfaces. These three degrees of freedom are not always separable from one another in practice, but all three are gauge-free (physical) dynamical variables. In contrast, the lapse function $N(x, y, z, t)$, together with the shift functions, merely determines how a specific time *coordinate* is related to this many-fingered time.

Therefore, the three-geometry $^{(3)}G$, representing the *dynamical state* of general relativity, is itself the 'carrier of information on physical time' (Baierlein, Sharp and Wheeler 1962): it *contains* physical time rather than *depending* on it. By means of the Einstein equations, $^{(3)}G$ determines a *continuum of physical clocks*, that is, all time-like distances from an 'initial' $^{(3)}G_0$ (provided a solution of the corresponding boundary value problem does exist). Given yesterday's geometry, today's geometry could not be tomorrow's – an absolutely non-trivial statement, since $^{(3)}G_0$ by itself is not a complete initial condition that would determine the solution of (5.7) up to a gauge. A mechanical clock can meaningfully go 'wrong'; for a rotating planet one would have to know the initial angle *and* the initial rotation velocity in order to read time from motion. However, a *speed* of three-geometry (in contrast to the *direction* of its velocity in superspace) would be as tautological as a 'speed of time'.

In this sense, Mach's principle (here with respect to time)[3] is anchored in general relativity: time must be realized by dynamical objects (such as spatial geometry). Dynamical laws that do *not* implicitly presume an absolute time are characterized by their *reparametrization invariance*, that is, invariance under monotonic transformations, $t \rightarrow t' = f(t)$. In general relativity, the time parameter t labels trajectories in superspace by the values of an appropriate time coordinate. No specific choice may then 'simplify' the laws according to Poincaré's definition (see Chap. 1), and no distinction between active and passive reparametrizations remains meaningful (see Norton 1989). It is therefore amazing to observe ongoing attempts to re-establish an external concept of time – even by means of 'phantom fields' (Thiemann 2006). The latter attempt was inspired (though not justified) by the problematic distinction between coordinate transformations and 'active' diffeomorphisms (see also Sect. 6.2.2).

Newton's equations are *not* invariant under a reparametrization. His time t is not an arbitrary parameter, but a dynamically preferred one ('absolute' time). Its reparametrization would merely be 'Kretzschmann invariant', that is, invariant under a trivial substitution of the old coordinates by new ones – thereby allowing for a reformulation of the dynamical laws by means of Coriolis-type pseudo-forces. Newton's equations can be brought into a reparametrization-invariant *form* only by artificially parametrizing the time variable t itself, $t(\lambda)$, and treating it as an additional dynamical variable with respect to λ. If $L(q, \dot{q})$ is the original Lagrangean, this leads to the *new*

---

[3] See Barbour and Pfister (1995) for various interpretations of Mach's principle.

variational principle

$$\delta \int \tilde{L}\left(q, \frac{\mathrm{d}q}{\mathrm{d}\lambda}, \frac{\mathrm{d}t}{\mathrm{d}\lambda}\right) \mathrm{d}\lambda := \delta \int L\left(q, \frac{\mathrm{d}q}{\mathrm{d}\lambda}\frac{\mathrm{d}\lambda}{\mathrm{d}t}\right) \frac{\mathrm{d}t}{\mathrm{d}\lambda}\mathrm{d}\lambda = 0 , \qquad (5.28)$$

where the absolute time $t(\lambda)$ has to be varied, too. This procedure also helps to understand the meaning of the 'Δ-variation' that often appears somewhat unmotivated in analytical mechanics (see Sect. 8.6 of Goldstein 1980). Evidently, (5.28) is invariant under the reparametrization $\lambda \to \lambda' = f(\lambda)$.

Eliminating the formal variable t from (5.28) then leads to *Jacobi's principle* (see below), which was partially motivated by the pragmatic requirements of astronomers who did not have better clocks than the objects they were dynamically describing. These clocks, which define *ephemeris time*, are given by stellar positions when compared with *tables of ephemeris* produced by colleague astronomers. Since all celestial motions must be more or less 'perturbed' by others, they do not offer any obvious way to define Newton's time operationally. Jacobi's principle allowed astronomers to solve the equations of motion without explicitly using Newton's time. Einstein's equations of general relativity, on the other hand, are invariant under reparametrization of their time coordinate, $t \to t' = f(t)$, without any further and artificial parametrization $t(\lambda)$. There is no longer any time beyond the many-fingered dynamical variable contained in $^{(3)}G$!

In (5.28), $\mathrm{d}t/\mathrm{d}\lambda =: N(\lambda)$ may be regarded as a Newtonian lapse function (the relation between absolute time and a time parameter). For a time-independent Lagrangean L, t then appears as a cyclic variable. Its canonical momentum, $p_t := \partial \tilde{L}/\partial N = L - \sum p_i \dot{q}_i = -H$, which is conserved, is remarkable only because its quantization leads to the time-dependent Schrödinger equation. However, the 'super-Hamiltonian' $\tilde{H}$ that describes the extended system which includes $t(\lambda)$ is trivial:

$$\tilde{H} := \sum p_i \frac{\mathrm{d}q_i}{\mathrm{d}\lambda} + p_t \frac{\mathrm{d}t}{\mathrm{d}\lambda} - \tilde{L} = N\left(\sum p_i \frac{\mathrm{d}q_i}{\mathrm{d}t} - H - L\right) \equiv 0 . \qquad (5.29)$$

More dynamical content can be extracted from Dirac's procedure of treating $N(\lambda)$ rather than $t(\lambda)$ as a new variable. The corresponding momentum, $p_N := \partial \tilde{L}/\partial(\mathrm{d}N/\mathrm{d}\lambda) \equiv 0$, has to be regarded as a constraint, while the *new* super-Hamiltonian is

$$H_S := \sum p_i \frac{\mathrm{d}q_i}{\mathrm{d}\lambda} + p_N \frac{\mathrm{d}N}{\mathrm{d}\lambda} - \tilde{L} = NH . \qquad (5.30)$$

Although $\mathrm{d}N/\mathrm{d}\lambda$ cannot be eliminated in the usual way here by inverting the definition of canonical momentum $p_N(N, \mathrm{d}N/\mathrm{d}\lambda, \dots)$, it drops out everywhere in the Hamiltonian equations except in the derivative $\partial H_S/\partial p_N$, since it occurs only as a factor multiplying the vanishing p_N. The two new Hamiltonian equations related to the variable $N(\lambda)$ are (1) $\mathrm{d}N/\mathrm{d}\lambda = \partial H_S/\partial p_N = \mathrm{d}N/\mathrm{d}\lambda$, which is an identity, and (2) $\mathrm{d}p_N/\mathrm{d}\lambda = -\partial H_S/\partial N = -H$. Because $p_N \equiv 0$, one obtains the (secondary) *Hamiltonian constraint* $H = 0$ (but not $\equiv 0$), characteristic of reparametrization invariant theories. This result is the origin of the vanishing energy in (5.21), and will turn out to be important for quantum gravity. In general relativity, there are also three *momentum constraints*, characterizing invariance under spatial coordinate transformations, and related to the shift functions when chosen as formal dynamical variables.

Hamilton's new principle (5.28) can be written in the form

$$\delta \int \left(\sum p_i \dot{q}_i - H \right) \frac{\mathrm{d}t}{\mathrm{d}\lambda} \mathrm{d}\lambda = 0 .$$

For fixed energy value, $H = E$, the second term would cancel under this variation because of the new boundary conditions $\delta t(\lambda) = 0$. For the usual quadratic form of the kinetic energy, $2T = \sum a_{ij} \dot{q}_i \dot{q}_j = \sum p_i \dot{q}_i = 2(E - V)$, the integrand can in this case be written homogeneously *linear* in $\mathrm{d}q_i/\mathrm{d}\lambda$:

$$\delta \int \sqrt{2(E - V) \sum a_{ij} \frac{\mathrm{d}q_i}{\mathrm{d}\lambda} \frac{\mathrm{d}q_j}{\mathrm{d}\lambda}} \, \mathrm{d}\lambda = 0 . \tag{5.31}$$

This is Jacobi's principle (see Lanczos 1970), useful for fixed energy. It is manifestly invariant under reparametrization of λ, and can thus describe only timeless orbits $q_i(\lambda)$. Even though these nonrelativistic equations of motion could be explicitly simplified by using Newton's time, (5.31) evidently does not depend on the choice of λ.

In Newton's theory, the energy E depends on absolute velocities $\mathrm{d}q_i/\mathrm{d}t$. Jacobi's principle would therefore describe a 'Machian' theory only if the fixed energy represented a *universal* constraint. Barbour and Bertotti (1982) were able to propose an illuminating nonrelativistic toy model for Machian mechanics by means of the action principle

$$\delta \int \sqrt{-VT} \mathrm{d}t = 0 , \tag{5.32}$$

inspired by (5.31). It is universally invariant under reparametrizations of t (just like general relativity). Nothing new could then be obtained from parametrizing t in order to vary $t(\lambda)$ as in (5.28). Barbour and Bertotti also eliminated absolute rotations from their configuration space. While this has other important consequences, it is irrelevant for the problem of time. In general relativity, this 'Leibniz group', consisting of time reparametrizations and spatial rotations, would have to be generalized to the whole group of *diffeomorphisms* (general coordinate transformations). In order to eliminate any absolute meaning of a time *coordinate* on spacetime, the Hamiltonian constraint has to be understood as a local condition on the Hamiltonian *density*, since in field theory spatial coordinates serve as 'indices' – not as variables.

Barbour (1999) refers to the absence of a physically meaningful function $t(\lambda)$ in general relativity as its *timelessness*. However, parametrizable trajectories still permit asymmetric boundary conditions, which would define a direction of intrinsic time. This is different in *quantum* cosmology, where the Hamiltonian constraint, combined with the time–energy uncertainty relation, leads to a complete elimination of time (Sect. 6.2). In classical general relativity, even a constrained Hamiltonian would define trajectories which represent cosmic histories in the form of spacetime foliations that can be parametrized, although no *external* time is required for this purpose. While the global states which form these histories depend on this arbitrary foliation, the resulting spacetime geometry does not. So it defines an invariant many-fingered time, that is, all *proper times*, for all local objects, such as 'test clocks' or observers, uniquely.

In the Friedmann model (5.20), where the shift function has been chosen as $N \equiv 1$, the increment of the time coordinate t is identical (up to a sign) with the increment of proper times τ of 'comoving' matter (being at rest in the Friedmann coordinates, which fulfill the condition $N_i \equiv 0$). Elimination of the global time parameter t would here merely reproduce the equation of state $\rho(a)$ as the corresponding 'trajectory', since ρ is not an independent dynamical variable. There is evidently no intrinsic distinction between expansion and contraction of this 'universe'. The single variable a would determine proper times τ for comoving matter up to this ambiguity, since $\dot{a}^2$ is given as a function of a by the energy constraint (5.21).

Even for the exactly symmetric Friedmann universe, matter can be described *dynamically* by means of a homogeneous scalar field $\Phi(t)$. Its energy density may be chosen as

$$\rho = \frac{1}{2}(\dot{\Phi}^2 + m^2\Phi^2) \, . \tag{5.33}$$

The Hamiltonian of this simple 'quantum mechanical' model with respect to the variables $\alpha = \ln a$ and Φ, derived from (5.22) without cosmological constant, then reads

$$H = \frac{e^{-3\alpha}}{2} \left(p_\alpha^2 - p_\Phi^2 + ke^{4\alpha} - m^2\Phi^2 e^{6\alpha} \right) \, , \tag{5.34}$$

where the canonical momenta are $p_\alpha = e^{3\alpha}\dot{\alpha}$ and $p_\Phi = -e^{3\alpha}\dot{\Phi}$. A 'timeless orbit' for a closed universe $(k = 1)$ in this model is depicted in Fig. 5.7. The freely chosen initial field $\Phi(a_0)$ at some small value a_0 first decays with increasing a, before it enters the 'matter-dominated' era, where it oscillates about the a-axis until it reaches a turning point in a as a consequence of the assumed positive curvature.

In the case of a Hamiltonian constraint, $H(p, q) = 0$, multiplying the Hamiltonian by a function $f(p, q)$, that is, $H \rightarrow H' = fH = 0$, would only induce an orbit-dependent reparametrization $t \rightarrow t'(t)$. This is given by $dt'/dt = f(p(t), q(t))$, as can be seen by writing down the new Hamiltonian equations. For example, the choice $f \equiv -1$ would induce an inversion of the Hamiltonian time parameter for all trajectories. Therefore, the factor $e^{-3\alpha}$ in (5.34) is irrelevant for the timeless orbits and can be omitted.

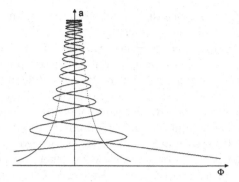

Fig. 5.7. Timeless classical orbit describing an expanding and recontracting dynamical Friedmann universe in terms of its expansion parameter a and a homogeneous massive scalar field Φ. *Dotted curves* represent vanishing Friedmann potential V as defined by (5.34). For slightly larger initial values $\Phi(a_0)$ than chosen in the figure, the 'inflation era', defined by the decaying initial field, would last over many orders of magnitude in a before the orbit entered the 'matter-dominated' era, where it performs a huge number of oscillations before reaching its turning point $a_{\max}$. (After Hawking and Wu 1985.) This dynamical description is very different in quantum gravity (see Fig. 6.3)

While this simple dynamical model cannot describe any thermodynamical aspects, it can be generalized by means of a multipole expansion on the Friedmann sphere,

$$\Phi(\chi, \theta, \phi, t) = \sum a_{nlm}(t) Q_{lm}^n(\chi, \theta, \phi) , \qquad (5.35)$$

where $Q_{lm}^n(\chi, \theta, \phi)$ are spherical harmonics on a three-sphere (Halliwell and Hawking 1985). The variable $\Phi(t)$ in (5.34) represents the monopole component, $\Phi = a_{000}$, since $Q_{00}^0 = 1$. A similar expansion of the metric tensor field g_{kl} requires vector and tensor harmonics in addition to the scalar harmonics Q_{lm}^n. Only the tensor harmonics turn out to represent physical (geometric) properties, while all others describe gauge degrees of freedom. In this 'perturbed Friedmann model', the time parameter t no longer automatically represents proper time on comoving world lines.

In (5.34) and its generalization to a multipole expansion, the kinetic energy of matter occurs with a negative sign (that is, with negative dynamical mass), since it entered the Hamiltonian as a source of gravity (representing negative potential energy). In Friedmann-type models, all gauge-free geometric degrees of freedom but the global expansion parameter a (or its logarithm) share this property (Giulini and Kiefer 1994, Giulini 1995), because gravitational waves imposed on a flat spacetime possess gravitating positive energy. The kinetic energy is thus not positive definite in cosmology, while the metric in infinite-dimensional superspace that it defines by its quadratic form is *super-Lorentzian* (with signature $+---\ldots$).[4]

This fact has important consequences. In the familiar case of mechanics, vanishing kinetic energy, $E - V = 0$, describes turning points of the motion. However, since there are no forbidden regions for indefinite kinetic energy, the boundary $V = V - E = 0$ does not force the trajectories to come to a halt and reverse direction here. Rather, this condition now describes a smooth transition between 'subluminal' and 'superluminal' directions in superspace (not in space!), as can be seen in Fig. 5.7. A trajectory would be reflected from an *infinite* potential 'barrier' only if this were either negative at a time-like boundary, or positive at a space-like one. Reversal of the cosmic expansion at $a_{\max}$ requires the vanishing of an appropriate $V_{\mathrm{eff}}(\alpha)$ that includes the actual kinetic energy of the other degrees of freedom (similar to the effective radial potential in the Kepler problem). It is evident that this behavior must be important for a reversal of time and its arrow.

In the Friedmann model, a point on the trajectory in configuration space determines Friedmann time t (that could be read from comoving test clocks) – except where the curve intersects itself. In a mini-superspace with more than two degrees of freedom (adding a material clock, for example), physical time on a trajectory is generically *unique*, since intersections could occur only accidentally. This demonstrates that the essential requirement for the state to represent a carrier of information about time is reparametrization invariance of the dynamical laws – not its spacetime-geometric interpretation.

Although a time parameter is in general physically meaningless in these theories, it is often misused for an inappropriate interpretation. An example is Veneziano's (1991) string model, based on a dilaton field Φ. Its equations of motion lead to a time dependence of the form $f(t - t_0)$, with an integration constant t_0 that determines the value of the time parameter at the big bang (where $\alpha = -\infty$). A translation $t_0 \to t_0 + T$ would thus be meaningless (as already pointed out by Leibniz). The solution for $t < t_0$, where expansion *accelerates* exponentially in this model, has been interpreted as 'pre-big bang', while the absence of a smooth connection between pre- and post-big bang has been called a 'graceful exit problem' (Brustein and Veneziano 1994). However, this mathematical model has simply two different solutions, which could conceivably be related through an infinite parameter time, $t = \pm\infty$ – similar to Schwarzschild time at a horizon. Coordinate times $t < t_0$ would then represent physical times *later* than $t > t_0$, while a continuation through t_0 is merely formal (Dabrowski and Kiefer 1997).

---

[4] There is also a *local*, 6-dimensional Lorentzian metric in superspace, corresponding to the 6 degrees of freedom of the submatrix g_{kl} at every space point, such that there seems to be an infinity of time-like variables (see Sect. 6.2.2). However, all but one of them are unphysical gauge degrees of freedom in a Friedmann type universe.

The shift functions N_i of (5.27) can be chosen to vanish even when spatial symmetries are absent. The secondary *momentum constraints* $H_i := \partial\tilde{H}/\partial N_i = 0$, which warrant conservation of vanishing canonical momenta p_{N_i}, and which are fulfilled automatically for the Friedmann solution because of its symmetry, then have to be solved explicitly. The lapse function $N(x, y, z, t)$ now determines genuine *many-fingered* time (as a spatial *field* on the dynamically evolving hypersurface) with respect to the coordinate t. If N is nonetheless chosen as a function of t alone, the foliation proceeds everywhere according to physical time (*normal* to the hypersurface, with fixed 'comoving' coordinates).

This may not always be a convenient choice. For example, observers coming very close to a black hole horizon would observe the stars moving very fast through a little hole that remains in the sky above the horizon because of their extreme time dilation. In a universe that is bound to recontract they could reach the contraction era within very short proper times. This renders the immediate vicinity of horizons very sensitive to a conceivable cosmic *final* condition, which may even exclude black hole horizons and singularities (see Zeh 1983, 2005a, and Sect. 6.2.3). In this case, a foliation according to *York time*, mentioned in Sect. 5.1, may be preferable, since it arrives 'simultaneously' at all final singularities. Note, however, that the external curvature scalar K, which defines York time, is *not* a function of state, $f(^{(3)}G)$.

Among the simplest inhomogeneous models are the spherically symmetric ones, with a metric

$$ds^2 = -N(\chi, t)^2 dt^2 + L(\chi, t)^2 d\chi^2 + R(\chi, t)^2 (d\theta^2 + \sin^2\theta \, d\phi^2) \, . \quad (5.36)$$

They contain one remaining spatial gauge function, that has to be eliminated by means of the momentum constraint $H_\chi = 0$. This is analogous to Gauß's law in electrodynamics, as it similarly refers to the radial coordinate.

Qadir (1988) proposed an illustrative toy model for such an inhomogeneous universe (Fig. 5.8). It forms a generalization of the Oppenheimer–Snyder model for the gravitational collapse of a homogeneous spherical dust cloud (see Misner, Thorne and Wheeler 1973, Chap. 32). The latter model pastes (or 'sutures') a comoving spherical surface surrounding part of a contracting closed Friedmann solution (representing the dust cloud) consistently to the external region a Schwarzschild–Kruskal solution. Qadir then pastes this Schwarzschild solution in turn to another (much larger) partial Friedmann solution with much smaller energy density (his universe proper). This pasting at two spatial boundaries, with Friedmann radial coordinate values χ_1 and χ_2, say, is consistent only if the total masses of the two partial Friedmann universes are identical, and can thus be identified with the Schwarzschild mass M characterizing the partial vacuum solution. The latter forms a strip from Fig. 5.2 between two non-intersecting geodesics that lead from the past to the future Kruskal singularity (big bang and big crunch).

In order to comply with the Weyl tensor hypothesis as much as possible, Qadir assumed the 'Schwarzschild corridor' to be absent at the big bang.

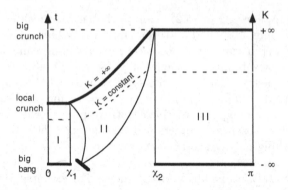

Fig. 5.8. Qadir's 'suture model' of a collapsing homogeneous dust cloud, I, as part of an expanding and recontracting Friedmann universe, III. The Friedmann spheres at χ_1 (*left*) and χ_2 (*right*) are initially identified. The Weyl tensor, representing the gravitational degrees of freedom, is thus chosen to vanish initially (except at the spatial boundary between the two regions I and III), but will grow by means of an emerging 'Schwarzschild–Kruskal corridor', II, (a strip from Fig. 5.2). The spatial boundaries of the three spacetime regions have to be identified (including proper times on them, all chosen to start at the big bang). According to a picture due to Penrose, the singularity inside the black hole (region I) together with its attached Kruskal singularity (in region II) appears as a 'stalactite' hanging from the 'ceiling' (which represents the big crunch singularity in region III). In contrast, there is only one (piecewise homogeneous) big bang singularity (a flat floor in Penrose's picture) at $K = -\infty$, that is chosen as the first slice of the foliation (corresponding to $t = 0$)

The density discontinuity then represents an initial inhomogeneity. Since the denser part of this toy universe feels stronger gravitational attraction than the less dense one, its expansion decelerates (or its contraction accelerates) faster. A vacuum corridor must then form and grow in size with increasing temporal distance from the big bang. As the energy–momentum tensor vanishes in the Schwarzschild–Kruskal region, the curvature is there entirely due to the Weyl tensor, while the latter vanishes inside the two partial Friedmann universes. The time arrow of this process of 'gravitational monopole radiation' (the formation of the corridor with its non-zero gravitational degrees of freedom) is once again a consequence of the special initial condition.

This model is certainly interesting as an illustration of the Weyl tensor hypothesis, but it does not describe statistical (entropic) aspects. For this purpose, many multipoles of (5.35) would have to be taken into account as radiation. Qadir's cosmic evolution process simply describes an example of motion away from the chosen initial state – similar to what is normally found in unbound mechanical systems regardless of any statistical considerations.

General Literature: Chap. 21 of Misner, Thorne and Wheeler 1973; Barbour 1999; Kiefer 2007.

6

The Time Arrow in Quantum Cosmology

> Our mistake is not that we take our theories
> too seriously, but that we do not take them
> seriously enough. (Stephen Weinberg 1977)
> – Well, but *which* theories?

The founders of quantum theory invented their theory as a theory of atoms, that was soon successfully applied also to other microscopic systems. Macroscopic objects were thought to require the established classical concepts even though they consist of atoms. This hardly consistent traditional point of view (that would also exclude quantum cosmology) seems to be slowly changing under the impact of more recent interpretations, which allow one to describe the world in terms of a universally valid quantum theory (Sect. 4.6).

Another obstacle to quantum cosmology is that a description of the whole Universe seems to require a 'theory of everything', which is elusive. While there are various mathematically deep and physically even plausible proposals for such a theory, physics is an empirical science. Physical cosmology should therefore only extrapolate empirically founded concepts and laws. Mathematical cosmological models may be important and interesting in their own right, and some of them may prove physically successful in the future, but reality has usually offered great conceptual surprises that could not have been foreseen by mathematical reasoning or pure logic.

Physical cosmology should not therefore rely on any details of unconfirmed unified quantum field theories, for example. Only the general framework of quantum theory may be regarded as empirically sufficiently founded to draw cosmological conclusions from it. This framework includes, first of all, the superposition principle and the unitarity of dynamics (in other words, a general wave function and a Schrödinger equation). In cosmology, this requires an answer to the fundamental problem of what quantum theory *means* in the absence of external observers or measurement devices. Physical cosmology must therefore depend on the interpretation of quantum theory (as discussed in Sect. 4.6) in an essential way. A pragmatic probability interpretation with respect to external observers is obviously ruled out, since the very concept of cosmology presumes an objective (though in principle hypothetical) reality. Quantum *field* theory has instead traditionally been used and confirmed as a method for calculating S-matrix elements, which describe probabilities for scattering events. This amounts to applying a collapse of the wave function after each elementary scattering process, and it would be insufficient for consistently describing objects which make up the Universe, such as condensed matter, complex systems (including measurement devices and observers), macroscopic fields, and global spacetime structure.

The general quantum framework is usually applied in the form of a 'quantization' of a classical theory (see Sect. 4.1.1) – in particular of the mechanics of particles, which are kinematically described as space points. By quantization I mean here[1] the application of the superposition principle to the elements of a classical configuration space (thus defining a wave function on it), and the construction of the corresponding quantum Hamiltonian by replacing variables and their canonical momenta by operators acting on wave functions. The second part is ambiguous because of the factor ordering problem.

We can now re-interpret this quantization procedure as the conceptual reversal of a physical decoherence process that led to the classical appearance of the system under consideration. This explains why this quantization cannot be expected to define a unique result, but requires further empirical input. The quantization of many-particle mechanics leads non-relativistically 'back' to a consistent and successful quantum theory: quantum mechanics. Some other 'particle' properties (such as spin or isotopic spin) have *no* similar classical correspondence. The quantization of classical *fields* in this canonical way leads to wave functionals on the configuration space for field amplitudes. It does not in general directly define a consistent quantum theory, although it can often be rendered consistent by a mere renormalization of its fundamental parameters. This is evidence that a fundamental quantum theory may be quite independent of any classical theory that could be quantized in this way. For example, relativistic quantum mechanics led to the discovery that field amplitudes of *not classically observed* fermion fields rather than particle positions define the correct arena for the wave function(al) – an approach that is somewhat misleadingly called a 'second quantization', since the fermion fields were first discovered as effective 'single-particle wave functions' (see Zeh 2003). The underlying fields (on space) define a local *basis* (the 'stage' for quantum

---

[1] This interpretation is quite different from the original and literal meaning of the term 'quantization' as a *discretization* of certain quantities. For example, 'light quanta' can be understood as a *consequence* of the eigenvalue problem in terms of wave functions for the amplitudes of free field modes, dynamically described as harmonic oscillators. These fundamental aspects of quantum theory are often hidden behind a collection of recipes to perform calculations (such as perturbation theory in terms of Feynman graphs). In particular, a 'quantization of time' (Sect. 6.2) does *not* require a quantum of time – just as the quantization of particle motion does not require a quantum of length (or a spatial lattice).

dynamics) that spans the required Hilbert space. This structure permits the formulation of *local dynamics* by means of a Hamiltonian density in spite of generically nonlocal states. It may therefore be useful – though also dangerous and certainly insufficient – to investigate mathematical models for a unified quantum field theory solely by investigating certain *classical* fields on three- or higher-dimensional spaces, rather than consistently taking into account their quantum nature from the beginning (for instance in terms of wave functionals of these fields as representing the true quantum reality).

Extrapolating unitary dynamics to the whole Universe requires an Everett type interpretation (see Sect. 4.6). Hugh Everett (1957) seems to have first seriously considered a wave function of the Universe,[2] that must then include *internal observers*. Although he may have had in mind the quantization of general relativity with its cosmological aspects, Everett applied his ideas, which were based on a time-dependent Schrödinger equation, to nonrelativistic quantum theory. His main interpretational obstacle was the entanglement arising from measurements described by means of von Neumann's unitary interaction (4.32). This led him to his 'extravagant' interpretation (in Bell's words) in terms of *many* quasi-classical 'branches' of the world, which are separately experienced, but are all assumed to exist in one superposition that defines the true and dynamically consistent quantum world. Beyond measurements proper and occasional interactions he does not seem to have regarded entanglement as particularly important (see Tegmark 1998).

The quantitative considerations reviewed in Sect. 4.3 demonstrate that *uncontrollable* 'measurement-like' interactions with the environment are essential and unavoidable for almost all systems under all realistic circumstances. Strong entanglement is, therefore, a *generic* aspect of quantum theory. The more macroscopic a system, the stronger its entanglement with its environment. The concept of a (pure) quantum state can be consistently applied only to the Universe as a whole (Zeh 1970, Gell-Mann and Hartle 1990). This seems to be a far more powerful argument for the need of quantum cosmology than an attempt to construct a unified quantum field theory.

The second pillar of physical cosmology is general relativity. It is empirically confirmed only as a classical theory, but this fact can be well understood by decoherence again (see Sects. 4.3.5 and 6.2.2). Exactly classical gravity would lead to inconsistencies with the uncertainty principle. Applying the quantization rules to the Hamiltonian formalism of general relativity (described in Sect. 5.4) leads to a non-renormalizable 'effective' quantum gravity that cannot be exact, but may be expected to be appropriate as a low energy limit. This readily allows us to discuss a number of important novel conceptual problems that must come up, in particular the need for a 'quantization of time' (Sect. 6.2).

---

[2] Thibault Damour (2006) has recently presented evidence that Everett was originally stimulated by remarks Albert Einstein made about quantum theory during his last seminar, given at Princeton in 1954.

The quantum state of the Universe must therefore include gravitational degrees of freedom (entangled with matter) in an essential way. However, many quantum cosmological aspects may be formulated on a quasi-classical background spacetime, using a given foliation parametrized by a time coordinate t. Global states can then be dynamically described by means of a time-dependent Schrödinger equation with respect to this coordinate time t. This formalism will be *derived* from quantum gravity (with its quantized concept of an *intrinsic* time) in Sect. 6.2.2 as an approximation. *Global* states (such as those of quantum fields) depend on a foliation (or a reference frame) even on flat spacetime, while the density matrix of any *local* system should be invariant under a change of foliation that preserves its local rest frame – a requirement that does not seem to have attracted much attention.

If the Quantum Universe is thus conceptually regarded as a whole, it does not decohere, since there is no further environment. Decoherence is meaningful only for *subsystems* of the Universe (or for subsets of variables), and with respect to observations by other subsystems (internal 'observer-participators'). If no real collapse of the wave function is assumed to apply, one is then *forced* to accept Everett's global wave function, which describes a superposition of at least all 'possible' outcomes of measurements and measurement-like processes that ever occurred in the Universe. This global quantum state may always be assumed to be pure, since a global density matrix could be consistently understood as representing incomplete information about such a pure state. A measurement that merely selects a subset from those states which diagonalize this density matrix would be equivalent to a classical measurement (as depicted in Fig. 3.5 – in contrast to Fig. 4.3).

The decoherence of subsystems by their environment according to a *global* Schrödinger equation leads dynamically to robust Everett branches. They represent dynamically autonomous *components* of the global wave function, which may factorize in the form $\phi_{obs1}\phi_{obs2}\ldots\psi_{rest}$ with respect to 'observer states' that may describe objectivizable memory (see Sect. 4.3.2 and Tegmark 2000). This unitary evolution requires a fact-like arrow of time, corresponding to a cosmic initial condition of type (4.59). Branching into components which contain definite observer states has to be *taken into account* in addition to the unitary evolution as an *effective* dynamics in order to describe the history of the (quasi-classical) 'observed world' in quantum mechanical terms (see Sect. 4.6 and Fig. 4.3). However, this need *not* represent a modification of the fundamental dynamical laws, since this indeterminism affects the observer rather than the quantum world. The decrease of physical entropy characterizing the 'apparent collapse' experienced by the subjective observer may be negligible on a thermodynamical scale, and in comparison to the entropy increase by decoherence in the usual situation of a measurement. Yet it may have dramatic consequences for global phase transitions that describe a dynamical symmetry-breaking of the vacuum. This will now be discussed.

6.1 Phase Transitions of the Vacuum

Heisenberg (1957) and Nambu and Jona-Lasinio (1961) invented the concept of a vacuum that breaks symmetries of a fundamental Hamiltonian 'spontaneously' (in a fact-like way) – just as *most* actual states of physical systems do. This proposal was based on an analogy between the vacuum (the ground state of quantum field theory) and the phenomenological ground states of macroscopic systems, such as ferromagnets or solid bodies in general. Their asymmetric ground states lead to specific modes of excitation, which in quantum theory define quasi-particles (phonons, for example). The corresponding occupation number eigenstates span specific partial Hilbert spaces ('Fock spaces'). A symmetry-violating vacuum may similarly lead to *Goldstone bosons* or other collective modes, based on space-dependent oscillations of the order parameter about its macroscopic (collective) 'orientation' – see below.

A symmetry-breaking (quasi-classical) 'ground state' is in general not even an eigenstate of the fundamental (symmetric) Hamiltonian; it may only form an eigenstate of an effective (asymmetric) Fock space Hamiltonian. While non-diagonal elements of the exact Hamiltonian which connect states of different collective orientation of these many-body systems (lying in different Fock spaces), are usually extremely small, they would be essential to determine its exact eigenstates, since the diagonal elements for all states related by a symmetry transformation must be degenerate.

The symmetry-breaking vacuum was originally understood as part of the kinematics of a field theory, while the dynamics was then assumed to be completely defined by means of the Fock space Hamiltonian. Later, the analogy was generalized to allow for a *dynamical* phase transition of the vacuum during the early stages of the Universe. This may be induced by the variation of some global parameter (such as a rapid decrease of energy density, reflecting the expansion of the Universe). The arising 'unitarily inequivalent' different Fock spaces can then be interpreted as robust Everett branches or collapse components. Even the empirical P or CP-violating terms of the (effective) weak-interaction Hamiltonian may have *emerged* dynamically in this way by means of an apparent or genuine collapse of the wave function that led to a specific vacuum.

A popular model for describing symmetry-breaking in non-perturbative quantum field theory is the 'Mexican hat' or 'wine bottle potential' of the type $V(\Phi) = a|\Phi|^4 - b|\Phi|^2$ (with $a, b > 0$) for a fundamental complex matter field Φ (such as a *Higgs field*). It may possess a degenerate minimum on a circle in the complex plane, at $|\Phi| = \Phi_0 > 0$, say. The classical field configurations of lowest energy may then be written as $\Phi \equiv \Phi_0 e^{i\alpha}$, with an arbitrary phase α. They break the dynamical symmetry under rotations in the complex Φ-plane. These classical ground states correspond to different quantum mechanical vacuum states $|\alpha\rangle$ (for example described by narrow Gauß packets of α-eigenstates). One of them, $|\alpha_0\rangle$, say, is assumed to characterize our observed world (while the specific value of α_0 is in this case observationally meaningless).

A *dynamical* phase transition of the vacuum can now be described by assuming that the Universe was initially in the symmetric vacuum $|\Phi \equiv 0\rangle$. This may later become a 'false' vacuum (a *relative* minimum) through a change of the parameters a and b. The state of the observed universe is then assumed to undergo a transition into a specific Fock space vacuum $|\alpha_0\rangle$. If potential energy is thereby released in a 'slow roll' (similar to latent heat in a phase transition), it must be transformed into excitations (particle creation). Evidently, this symmetry-breaking process requires effective deviations from the Schrödinger equation – similar to a measurement process.

If the initial state is here assumed to be pure, a *unitary* evolution (similar to von Neumann's measurement) leads to a symmetric superposition of all asymmetric states. For example, the symmetric superposition of all Fock space vacua,

$$|0_{\text{sym}}\rangle = C \int |\alpha\rangle \mathrm{d}\alpha \neq |\Phi \equiv 0\rangle \,, \tag{6.1}$$

may possess an even lower energy expectation value than $|\alpha\rangle$, and may thus represent an approximation to the ground state of the full theory. A globally symmetric superposition of type (6.1) would persist even when its components on the RHS contain or develop uncontrollable excitations *in their Fock spaces*, while these components then form dynamically independent Everett branches. The superposition itself describes *intrinsic complexity*, but not a global asymmetry. If $\pi_\alpha := \mathrm{i}\partial/\partial\alpha$ generates a gauge transformation, (6.1) describes a state obeying a gauge constraint, $\pi_\alpha |\psi\rangle = 0$ (see Sect. 6.2).

Each homogeneous *classical* state α_0 would permit excitations in the form of small space-dependent oscillations, $\alpha_0 + \Delta\alpha(\boldsymbol{r}, t)$. Quantum mechanically they describe massless *Goldstone bosons* (excitations with vanishing energy in the limit of infinite wavelength because of the degeneracy). Their degrees of freedom are thus *created* by the intrinsic symmetry breaking, and their observation demonstrates that the collective variables (including corresponding 'gauge' degrees of freedom) do not describe mere redundancies. These new variables may be thermodynamically extremely relevant. So it is remarkable that the most important cosmic entropy capacities are represented by zero-mass bosons: electromagnetic and gravitational fields (Zeh 1986a, Joos 1987). These capacities are not only relevant for physical entropy (such as in the form of heat), but also for the formation of entanglement between different spatial regions. This seems to be important for the 'arrow of quantum causality' (Sect. 4.6).

In contrast to the false vacuum, the symmetric superposition (6.1) would already describe a nonlocal state. If one neglects Casimir–Unruh type correlations (see Sect. 5.2), each vacuum $|\alpha\rangle$ may be written as a direct product of vacua on volume elements ΔV_k,

$$|\alpha\rangle \approx \prod_k |\alpha\rangle_{\Delta V_k} \,. \tag{6.2}$$

This non-relativistic approximation describes a pure vacuum state on each volume element (local subsystem) ΔV_k, while the superposition (6.1) would lead to 'mixed states' for them:

$$\rho_{\Delta V_k} \propto \int |\alpha_{\Delta V_k}\rangle\langle\alpha_{\Delta V_k}| \, d\alpha \,, \qquad (6.3)$$

formally representing Zwanzig projections $\hat{P}_{sub}$. However, this density matrix would be meaningful only for an *external* observer of the global state (who could not live in one of the Fock spaces). It describes a canonical distribution of Goldstone bosons with infinite temperature (since then $e^{-E/kT} \to 1$). Therefore, only a (genuine or apparent) collapse into *one* component α_0 gives rise to the pure (cold and not entangled) vacuum (6.2) experienced by an internal local observer who lives in this Fock space.

Order parameters such as α may differ in different spatial regions (similar to Weiss regions of a ferromagnet). If these regions are macroscopic, and thus decohere to become 'real' (see Sect. 4.3.1), they break translational symmetry (Calzetta and Hu 1995, Kiefer, Polarski and Starobinsky 1998, Kiefer et al. 2006). This scenario has now become 'standard' in quantum cosmology – although its interpretation varies. A homogeneous superposition of entangled *microscopic* inhomogeneities would represent 'virtual' symmetry breaking (in classical language circumscribed as 'vacuum fluctuations').

6.2 Quantum Gravity and the Quantization of Time

> Um sie kein Ort, noch weniger eine Zeit;
> Von ihnen sprechen ist Verlegenheit.
> (Mephisto advising Faust to time travel)

The compatibility of general relativity and quantum theory has often been questioned. This seems to be a prejudice, that derives from various roots:

Einstein's attitude regarding quantum theory is well known. He is even claimed to have remarked that a quantization of general relativity would be 'childish' – although he also emphasized the importance of reconciling his theory with quantum theory. Another position holds that gravitons may be unobservable in practice, and the quantization of gravity hence *not required* (von Borzeszkowski and Treder 1988). However, a classical gravitational field or spacetime metric is *inconsistent* with quantum mechanics, since it would always allow one in principle to determine the exact energy of a quantum object – in conflict with the uncertainty relations. This has been known since the early Bohr–Einstein debate (see Jammer 1974, for example), while other consistency problems regarding an exactly classical spacetime metric were raised by Page and Geilker (1982). Concepts of quantum gravity will turn out to be essential for cosmology and the definition of a master arrow of time. The classical *appearance* of spacetime cannot be regarded as an argument against its quantization, since these classical aspects may be understood within a universal quantum theory in a similar way to all other quasi-classical properties (Sect. 4.3.5).

One often finds also arguments that the canonical quantization of general relativity does not lead to a renormalizable theory, and must therefore be wrong. This argument would apply if quantum gravity was assumed to be an exact theory. However, it can only be expected to represent an 'effective theory' that describes specific low energy aspects of an elusive unified field theory. We know that QED, too, has to be modified and replaced by electroweak theory at high energies, while it remains an excellent and consistent description of all relevant phenomena at low energies. Its most general quantum aspects (described in terms of QED wave functionals) are in fact observed for laser fields in cavities. An analogous (though technically and conceptually more demanding) canonical method of quantizing general relativity leads to the Wheeler–DeWitt equation (6.4) below (DeWitt 1967). Why should the Einstein equations be saved from quantization, while the Maxwell equations are not? The conceptual consequences of quantum gravity, in particular those for cosmology, have turned out to be profound even at this level of a low energy approximation.

The construction of a unified theory certainly represents the major challenge to quantum field theory at a fundamental level. Such a theory must become important in the vicinity of spacetime singularities (inside black holes or close to the big bang), but may also have cosmological consequences. In the absence of any observational confirmation, the latter have to be regarded as 'mathematical cosmology', that remains physically entirely speculative. Candidate models are often studied just as classical theories – sometimes including certain 'quantum corrections'. The surprising claim that *M-theory* may eventually lead to an *explanation* of quantum theory (Witten 1997) seems to be based on an elementary misunderstanding of quantum mechanics and its empirical basis.

A second approach to overcome fundamental problems of quantum gravity is canonical loop quantum gravity (Ashtekhar 1987, Rovelli and Smolin 1990, Thiemann 2006b, Nicolai, Peeters, and Zamaklar 2005). As it does not necessarily require a unification with other field theories, it does not contain most of the speculative elements of string theories, for example. To some extent it may be regarded as a specific though non-trivial renormalization of general relativity. Since this includes a radical formal redefinition of many phenomenological concepts, mostly by means of active diffeomorphisms that would severely affect also *classical* general relativity, it may help to find the correct configuration space on which the ultimate wave function may be defined. However, the relation of its quantization procedures (such as 'Bohr compactification' – invented by the mathematician Harald August Bohr) to the empirically founded quantization concepts must be regarded as highly questionable.

If quantization can indeed be understood as the conceptual reversal of a physical process of decoherence, the thereby recovered superpositions of the (effective) classical quantities should at least define an *effective* quantum gravity – similar to quantum mechanics, which is an effective theory in spite of more general quantum field theory. Therefore, the Wheeler–DeWitt equation in its field representation (here defined in terms of three-geometries) appears as the method of choice for 'physical' quantum cosmology. Questions of interpretation related to those for the wave function in general then seem to be more urgent at this stage than the consequences of speculative attempts to solve consistency problems that arise at high energies or in connection with a complete renormalization procedure.

A major problem that nonetheless prevents many physicists from accepting the Wheeler–DeWitt equation as appropriately describing quantum general relativity is the absence of any time parameter in the case of a closed universe (see Isham 1992). According to the Hamiltonian formulation reviewed in Sect. 5.4, one would naively expect free gravity to be described by a time-dependent wave functional on the configuration space of three-geometries, $\Psi[^{(3)}G, t]$, dynamically governed by a Schrödinger equation, $i\partial\Psi/\partial t = H\Psi$. However, there is no longer an external time parameter in a consistent quantum description, and nor are there *trajectories* of appropriate physical clock variables, which could give this time dependence an interpretation. Different three-geometries $^{(3)}G$ (classically the carriers of 'information' about *many-fingered* time – see Sect. 5.4) occur instead as *arguments* of these wave functionals. In the absence of parametrizable trajectories $^{(3)}G(t)$ – that is, of space-time geometries $^{(4)}G$, neither proper times nor global time coordinates are available. Therefore, it appears conceptually quite consistent (see Zeh 1984, 1986b, Barbour 1986) that the quantized form of the Hamiltonian constraint, $H\Psi = 0$,[3] completely removes any time parameter t from the wave function of a kinematically closed (though not necessarily finite) universe. This consequence must be expected to remain valid in reparametrizable unified theories – even after renormalization.

If matter is again represented by a single scalar field Φ on space-like hypersurfaces defined by their three-geometries $^{(3)}G$, the Wheeler–DeWitt equation assumes the general form

$$H\Psi[\Phi, {}^{(3)}G] = 0 \ . \tag{6.4}$$

---

[3] Only because of the (here quite inappropriate) Heisenberg picture in terms of particles is the equation $H\psi = E\psi$ usually called a *stationary* Schrödinger equation, while in wave mechanical terms it describes *static* solutions. Even 'vacuum fluctuations' represent static entanglement in the Schrödinger picture. Similarly, eigenvalues of the momentum operator are no more than *formally* analogous to classical momenta (which are defined as time derivatives). These conceptual subtleties will turn out to be essential for a consistent interpretation of the Wheeler–DeWitt equation.

Even though it represents dynamics, it does not describe a one-dimensional succession of states (or a history labelled by a parameter t). This is the natural quantum consequence of a classically missing *absolute time* (the absence of any preferred time parameter). In spite of the Hamiltonian constraint, the *classical* Hamiltonian equations would still define time-dependent (though reparametrizable) trajectories, which allow the unique (one-dimensional) ordering of states by means of physical clocks ('physical time'). While this does *not* apply to the dynamics (6.4) of quantum gravity any more, we shall see in Sect. 6.2.2 that one may approximately construct quasi-trajectories by means of a WKB approximation and using decoherence. Note that (6.4) describes the whole (Everett) quantum Universe, while branching components describing quasi-classical spacetimes would have to obey a stochastic quantum Langevin equation (see Sect. 4.6). However, the absence of *fundamental* trajectories in Hilbert space now leads to the problem of how to pose an 'initial' condition that would be able to explain the arrow of time that is already required for the irreversible process of decoherence, which is needed to justify the branching.

If time in a *closed mechanical* universe was according to Mach consistently defined by motion (as discussed in Chap. 1 regardless of general relativity), there could also be no meaningful time-dependent Schrödinger equation. Instead of an external or absolute time parameter t, one would have to refer to a physical clock variable u, say, that is part of this universe and has to be quantized, too. A time-dependent wave function $\psi(x,t)$ is thus replaced by an entangled wave function $\psi(x,u)$ (Peres 1980b, Page and Wootters 1983, Wootters 1984). In the conventional probability interpretation, $\psi(x,u)$ would describe a probability amplitude *for* 'physical time' u – not *at* a time u. Then why do we always observe states 'at' such a definite time rather than their superpositions? The answer is that one has to expect the relevant clock variable u to become quasi-classical for reasons explained in Sect. 4.3. For example, an assumed 'dust of test clocks' that measured proper times would according to (6.4) decohere any superposition of three-geometries which correspond to different intrinsic times. (This comes close to what really happens in our Universe – see Sect. 6.2.2.)

Equation (6.4) does not yet represent the Wheeler–DeWitt equation in a form that can be used. In practice, one has to represent the three-geometry $^{(3)}G$ by a metric $h_{kl}(x^1, x^2, x^3) = g_{kl}(x_0^0, x^1, x^2, x^3)$ with respect to a certain choice of coordinates (see Sect. 5.4). The wave functional $\Psi[h_{kl}]$ must then be invariant under spatial coordinate transformations. This is guaranteed by the three secondary momentum constraints, classically described as $H_i = 0$ (with $i = 1, 2, 3$). In their quantum mechanical form they must again be imposed as constraint operators acting on the wave function: $H_i\Psi[h_{kl}] = 0$, similar to the Hamiltonian constraint. If the momentum constraints are satisfied, the wave functional $\Psi[h_{kl}]$ represents a functional on three-geometries, $\Psi[^{(3)}G]$, only. This seems to require that the operators H_i commute in the weak sense, that is, their commutators must again define constraints. However, this may not be necessary if the effective theory lives in an Everett branch that breaks gauge symmetry – similar to an α-vacuum in (6.1). This would in turn require that the constraints define infinitesimal *physical* transformations – not just redundancies (see Giulini, Kiefer and Zeh 1995) as usually assumed for coordinate transformations. A similar conceptual problem occurs for many other gauge transformations.

General Literature: Kiefer 2007.

6.2.1 Quantization of the Friedmann Universe

A simple toy model of a quantum universe can be constructed by quantizing the classical Friedmann universe described in Sect. 5.3, presuming exact homogeneity and isotropy (Kaup and Vitello 1974, Blyth and Isham 1975). This leads to a reduced Wheeler–DeWitt equation for the two remaining variables, but in contrast to its classical counterpart it does not represent a reasonable approximation to reality. Symmetry requirements are much stronger in their quantum mechanical form than they are classically. For example, the rotation of a macroscopic spherical body produces a similar but microscopically different state, whereas a spherically symmetric *quantum* state is a symmetric superposition of all orientations, that would not be affected by this symmetry transformation any more. An exactly spherical quantum object therefore cannot possess any rotational degrees of freedom – compare the symmetric vacuum (6.1) or a superfluid in a spherical vessel. Rotational spectra are only found in the case of strong intrinsic symmetry breaking, as known for small molecules or deformed nuclei (Zeh 1967). Translations and rotations are thus *identity* operations when applied to a quantum Friedmann universe, which can therefore only be regarded as a very first step towards quantum cosmology (except perhaps very close to the big bang). The low-dimensional configuration space describing such a model is usually called a 'mini-superspace'.

If the Hamiltonian (5.34) is quantized in the usual way in its field representation, canonical momenta have to be replaced by the corresponding differential operators. In general, this leads to a factor-ordering problem for the Hamiltonian, since arbitrary terms proportional to the commutators $[p, q]$ could always be added before quantization. Although (5.34) looks quite 'normal', its straightforward translation into a Wheeler–DeWitt equation,

$$\frac{\mathrm{e}^{-3\alpha}}{2} \left(\frac{\partial^2}{\partial\alpha^2} - \frac{\partial^2}{\partial\Phi^2} - k\mathrm{e}^{4\alpha} + m^2\mathrm{e}^{6\alpha}\Phi^2 \right) \Psi(\alpha, \Phi) = 0 , \qquad (6.5)$$

is far from being trivial. For example, the result would have been different for quantization in terms of the expansion parameter a instead of its logarithm α. However, the choice used in (6.5) represents the invariant d'Alembertian with respect to DeWitt's 'superspace metric' that is defined by the quadratic form of momenta describing the kinetic energy – see (6.14) below. This specific factor- ordering is analogous to that for a point mass in flat space when formulated in terms of non-Cartesian coordinates. The prefactor $e^{-3\alpha}/2$ can be omitted from (6.5).

Even though the Wheeler–DeWitt equation is a stationary Schrödinger equation, it is of hyperbolic type – similar to a Klein–Gordon equation with variable mass – for reasons explained in Sect. 5.4. This fact offers the surprising possibility of formulating an *intrinsic initial value problem* in spite of the absence of any time parameter (see Sect. 2.1). The logarithmic expansion parameter α may be regarded as a time-like variable with respect to this intrinsic dynamics. The wave function on a 'space-like' hypersurface in superspace (for example at a fixed value of α) then defines an intrinsic *dynamical state* according to this dynamics. This intrinsic quantum dynamics with respect to the 'variable' α can also be written in the form

$$-\frac{\partial^2}{\partial \alpha^2}\Psi(\alpha,\Phi) = \left[-\frac{\partial^2}{\partial \Phi^2} + V(\alpha,\Phi)\right]\Psi(\alpha,\Phi) =: H_{\text{red}}^2\Psi(\alpha,\Phi)\,, \qquad (6.6)$$

in order to define a Klein–Gordon type *reduced Hamiltonian* H_{red}.[4] This dynamics is non-unitary, in particular as H_{red}^2 is not in general a non-negative operator. The 'reduced norm', $\int |\Psi(\alpha,\Phi)|^2 d\Phi$, is thus not generally conserved as a function of α. Although there is a conserved formal 'relativistic' two-current density in this mini-superspace,

$$j := \text{Im}\left(\Psi^* \nabla \Psi\right)\,, \qquad (6.7)$$

its direction depends on the sign of $i = \pm\sqrt{-1}$, which is physically meaningless in the absence of a time-dependent Schrödinger equation. There is in fact no reason even to expect a complex global solution of the real Wheeler–DeWitt equation.

The big bang and a conceivable big crunch would 'coincide' with respect to the intrinsic time variable α, while the expansion of the Universe becomes a tautology. The concept of a reversal of the cosmic expansion is an artifact of the classical description in terms of trajectories, such as in Fig. 5.7, while in more realistic models correlations of the expansion parameter α with quasi-classical physical clocks (including physiological ones) remain meaningful. So

---

[4] In *loop quantum gravity*, this differential equation has been replaced by a difference equation with respect to $p := a^2$. This *discrete* new variable can then be extended to negative values in a regular way (Bojowald 2003). This doubling of space would represent a space reflection, inverting the sign of the volume measure, such that the deterministic propagation through $p = 0$ (even in the continuous case) could be visualized as 'turning space inside out'. However, even if this extension of the concept of space could be vindicated in some way, the dynamical meaning of negative values of p (such as representing 'pre-big-bang times') had to be carefully analyzed. Since the low entropy 'initial' condition may be expected to apply at $p = 0$ rather than at $p = -\infty$ for reasons of symmetry, the physical direction of time would always point into the direction of growing $|p|$ – thus replacing the big bang by a reversal of the physical arrow of time in the past (see also Laguna 2006).

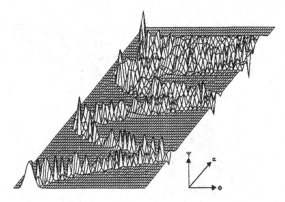

Fig. 6.1. Coherent 'wave tube' $\Psi(\alpha, \Phi)$ for the anisotropic indefinite harmonic oscillator (with $\omega_\Phi{:}\omega_\alpha = 7{:}1$) as a toy model of a periodically contracting and rebouncing quantum universe. It is here only plotted for the sector $\alpha > 0$ and $\Phi < 0$, since the solution is symmetric under reflections at both the α and the Φ axis (so wave tubes intersect at the right boundary). The intrinsic structure of the wave function is not completely resolved by the grid size used in this figure

what would 'happen' to a quasi-classical universe that were classically bound to recontract at some time?

One may discuss the quantum cosmological state in analogy to the 'stationary' wave function of a quantum 'particle' with fixed energy E, reflected from a spatial potential barrier (now a barrier in α). Since in timeless quantum gravity there is no reference phase $\mathrm{e}^{-i\omega t}$, one cannot distinguish between incoming and outgoing partial waves by their proportionality to $\mathrm{e}^{\pm ik\alpha}$.

Because of the hyperbolic nature of the Wheeler–DeWitt equation, narrow wave packets in Φ at fixed α lead to narrow 'wave tubes' that may approximately follow classical trajectories in mini-superspace (see Fig. 6.1). The case of positive spatial curvature, $k = +1$, is particularly illustrative. Its classical trajectories in mini-superspace would reverse direction with respect to α at some $\alpha_{\max}$ (Fig. 5.7). According to classical determinism, half of the trajectory (*defined* to represent the contracting universe) would be regarded as the dynamical successor of the other half (the expansion era). This deterministic relation is symmetric, since there is no absolute dynamical direction. The wave determinism described by the hyperbolic equation (6.6), on the other hand, propagates monotonically with α, and permits one to choose the whole initial condition (consisting of Ψ and $\partial\Psi/\partial\alpha$) on any 'space-like' hypersurface in superspace (e.g., at a small value of α). One could thus exclude precisely that part of the wave tube that would be required by classical determinism (Zeh 1988). How can these two forms of determinism (classical and quantum) be reconciled?

This dilemma can be resolved in analogy to conventional stationary states of quantum mechanics – though now including negative dynamical mass m_α

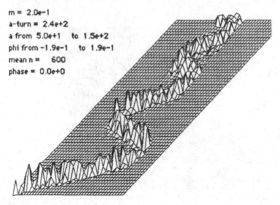

Fig. 6.2. Real-valued wave tube of the 'time'-dependent (damped) oscillator (6.5) in adiabatic approximation *without* taking into account the reflection at a_{max}, i.e., only the first term of (6.12) is used. Expansion parameter $a = e^\alpha$ is plotted upward and the amplitude of a homogeneous massive scalar field Φ from left to right

for the time-like variable. The simplest toy model for what is classically a periodically contracting and rebouncing universe is a free motion between reflecting boundaries of a narrow rectangle in α and Φ, where quantum mechanical solutions, such as real-valued wave tubes, can be constructed as superpositions of products of trigonometric functions matching the boundary conditions. (The potential barriers would have to be positive infinite for Φ-boundaries and negative infinite for α-boundaries.) In order to allow nontrivial zero-energy solutions, $H\Psi = 0$, the box lengths L_i have to be commensurable when accounting for the mass ratio, that is, $L_\alpha/L_\Phi = \sqrt{-m_\Phi/m_\alpha}\, k/l$, with integers k and l. Similar stationary wave tubes may be constructed in analogy to Schrödinger's coherent states from anisotropic harmonic oscillators (Fig. 6.1). In this case, one needs an indefinite potential, $V(\Phi, \alpha) = \left[(\omega_\Phi^2 \Phi^2 - \omega_\Phi) - (\omega_\alpha^2 \alpha^2 - \omega_\alpha)\right]/2$, where 'zero point energies' have been subtracted. In the commensurable case, now defined by $\omega_\alpha/\omega_\Phi = l/k$, solutions to the constraint $H\Psi = 0$ may be obtained as superpositions of the factorizing eigensolutions $\Theta_{n_\alpha}(\sqrt{\omega_\alpha}\alpha)\Theta_{n_\Phi}(\sqrt{\omega_\Phi}\Phi)$ of H, with eigenvalues $E = E_\alpha + E_\Phi = -n_\alpha \omega_\alpha + n_\Phi \omega_\Phi = 0$, in the form

$$\Psi(\alpha, \Phi) = \sum_n c_n \Theta_{nk}(\sqrt{\omega_\alpha}\alpha)\Theta_{nl}(\sqrt{\omega_\Phi}\Phi) \,. \tag{6.8}$$

If the coefficients c_n are chosen to define an 'initial' Gaussian wave packet in Φ at 'time' $\alpha = 0$, centered at some $\Phi_0 \neq 0$ and with $\partial\Psi/\partial\alpha = 0$, say, the resulting tube-like solutions propagate in α, following classical Lissajous figures in mini-superspace – just as for the conventional oscillator (DeWitt 1967).

In contrast to Schrödinger's time-dependent coherent states, which follow classical trajectories without changing their shape, these 'upside-down oscil-

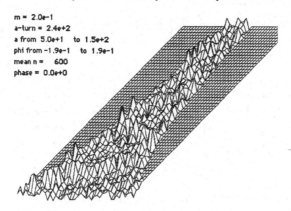

m = 2.0e-1
a-turn = 2.4e+2
a from 5.0e+1   to 1.5e+2
phi from -1.9e-1   to 1.9e-1
mean n =   600
phase = 0.0e+0

Fig. 6.3. Same as Fig. 6.2, but including the contribution of the reflected part: second term of (6.12). The coherent wave tube assumed to represent the expanding universe is here hardly recognizable against the background of the diffuse contribution representing the recontracting universe(s)

lators' display a rich intrinsic structure. Wave packets on different parts of a trajectory must also interfere with one another whenever they overlap – in particular close to the classical turning points. Interference between intersecting tubes would be suppressed by taking into account additional variables (higher-dimensional configuration spaces), since a projection onto mini-superspace – that is, tracing out all other variables – represents decoherence (see Sect. 6.2.2).

All components of (6.8) satisfy the usual boundary condition of normalizability in Φ and α. This choice is responsible for the reflection of wave tubes at the potential barriers. Although unusual for a conventional time parameter, it is consistent with the role of α as a dynamical variable. Such boundary conditions (if applied to the complete Quantum Universe) might even determine the solution of the Wheeler–DeWitt equation uniquely – provided there is a solution for its fixed zero eigenvalue at all. The degeneracy of the oscillator model (6.7), which allows the choice of 'initial' narrow wave packets, is evidently pathological. Narrow wave tubes can in general only be expected to arise as robust *branches* of the complete solution. They may not have to obey the boundary conditions individually.

An approximate solution can also be constructed for the Wheeler–DeWitt equation with a Friedmann Hamiltonian (6.5) – see Fig. 6.2 (Kiefer 1988). The oscillator potential with respect to Φ may here be assumed to be weakly α-dependent over many classical oscillations in Φ (shown in Fig. 5.7) except for small values of α. If α-dependent oscillator wave functions $\Theta_n(\Phi)$ – similar to those used in (6.8) – are now defined by the eigenvalue equation

$$\left(-\frac{\partial^2}{\partial \Phi^2} + m^2 e^{6\alpha}\Phi^2\right)\Theta_n\left(\sqrt{me^{3\alpha}}\Phi\right) = (2n+1)me^{3\alpha}\Theta_n\left(\sqrt{me^{3\alpha}}\Phi\right) \ , \quad (6.9)$$

one may expand a solution of (6.5) in terms of them as

$$\Psi(\alpha, \Phi) = \sum_n c_n(\alpha) \Theta_n(\sqrt{m} e^{3\alpha} \Phi) \ . \tag{6.10}$$

In the adiabatic approximation with respect to α (that may here be based on the Born–Oppenheimer expansion in terms of the inverse Planck mass – see Banks 1985 and Sect. 6.2.2), this leads to decoupled equations for the coefficients $c_n(\alpha)$:

$$\left[+\frac{\partial^2}{\partial \alpha^2} + 2E_n(\alpha) \right] c_n(\alpha) = 0 \ . \tag{6.11}$$

For positive spatial curvature, $k = 1$, the effective potentials

$$2E_n(\alpha) := (2n+1)m e^{3\alpha} - k e^{4\alpha}$$

become negative for $\alpha \to +\infty$. Even though $V(\alpha, \Phi)$ is positive almost everywhere in this limit - see (5.35), the Φ-oscillations are drawn into the narrow region $V < 0$ (in the vicinity of the α-axis – see Fig. 5.7) by damping.

So if one requires square integrability for $\alpha \to \infty$, only the exponentially decreasing partial wave solutions of (6.11) are admitted. Wave packets in Φ consisting of many oscillator eigenstates with quantum numbers $n \approx n_0$, say, may then be used to form wave tubes in α and Φ following the classical paths of Fig. 5.7 – see Fig. 6.2. However, in the Friedmann model, wave tubes cannot remain narrow wave packets in Φ when reflected at $a_{\max} = e^{\alpha_{\max}}$, since the turning point of the nth partial wave, $a_{\max,n} = (2n+1)m$, depends strongly on n. For values of a sufficiently below $a_{\max,n}$, the coefficients $c_n(\alpha)$ can according to Kiefer be written by means of a 'scattering' phase shift (caused by the reflection) in the form of a sum of incoming and outgoing (though real) waves. In the lowest WKB approximation one obtains ('asymptotically' in this sense)

$$c_n(\alpha) \propto \cos\left[\phi_n(\alpha) + n\Delta\phi\right] + \cos\left[\phi_n(\alpha) - n\Delta\phi + \frac{\pi}{4}a_n{}^2\right] \ . \tag{6.12}$$

Here,

$$\phi_n(\alpha) := \left(\frac{a_n}{4} - \frac{a}{2}\right)\sqrt{a(a_n - a)} + \left[\arcsin\left(1 - \frac{2a}{a_n}\right) - \frac{\pi}{2}\right]\frac{a_n^2}{8} - \frac{\pi}{4} \tag{6.13}$$

is a function of α and α_n, while the integration constant $\Delta\phi$ is the phase of the corresponding classical Φ-oscillation at its turning point in α. If coefficients are chosen, when substituting (6.12) into (6.10), such that the first cosines describe a (narrow) coherent oscillator wave packet, the 'scattering phase shifts' $\pi a_n{}^2/4$ of the second cosine terms cause the reflected wave to spread widely (see Fig. 6.3). Only for pathological potentials, such as the indefinite harmonic oscillator (6.8), or for integer values of $m^2/2$ in the specific model (6.4), can the phase shift differences be omitted as multiples of π.

Therefore, even the WKB approximation, which would suppress any dispersion of the wave packet, cannot in general describe an expanding and recollapsing quasi-classical universe by means of ('initially' prepared) wave packets that propagate as narrow tubes beyond the turning point. The concept of a universe deterministically expanding and recollapsing along a certain trajectory in superspace is as incompatible with quantum cosmology as the concept of an electron orbit in the hydrogen atom is with quantum mechanics. Many quasi-trajectories (wave tubes) describing expanding universes have to be superposed in order to obtain one quasi-classical contraction era (and vice versa). Decoherence has to select very different superpositions of partial waves $c_n \Theta_n$ in (6.10) to define robust branches in opposite eras. The reflection at a_{max} describes a quasi-stochastic quantum process – just as in a quantum scattering event. Compatibility problems for boundary conditions can thus affect only the total Wheeler–DeWitt wave function (the superposition of *all* branches). They would *not at all* occur for a non-normalizable Wheeler–DeWitt wave function that represents forever expanding universes ($k = 0$ or -1 in the case of $\Lambda = 0$).

All these simple models are far from being realistic. They are not only unable to describe statistical aspects or decoherence – they also neglect the important coupling between cosmic degrees of freedom and microscopic ones. In particular, the latter's ground states ('zero point motion') must in general depend on α and Φ. The cosmic variables are then subject to extreme decoherence (Barvinsky et al. 1999). Simplified models, as discussed above, may nonetheless appropriately describe certain important *conceptual* aspects of quantum cosmology.

General Literature: Ryan 1972, Kiefer 1988.

6.2.2 The Emergence of Classical Time

> If classical time emerges, it cannot emerge
> in classical time

A we have seen in Sect. 5.4, there is no dynamically preferred time parameter in general relativity or other reparametrization invariant classical theories. Moreover, the dynamical succession of global states, which may be conveniently described by means of a time coordinate, depends on the choice of a foliation. The resulting invariant spacetime geometry nonetheless defines many-fingered *physical* time, that is, absolute proper times as local controllers of motion, while any foliation represents a trajectory in superspace (a global history) that may then also be parametrized.

In *quantum* gravity, no global time parameter is generally available any longer, since there are no trajectories (that is, no one-dimensional successions of classical states with their physical clocks). Two given three-geometries are then not dynamically connected by a definite four-geometry (a spacetime).

Neither world lines nor their proper times, which could control Schrödinger or master equations for *local* systems, are defined, and three-geometry is no longer a reliable 'carrier of information about time'.

As explained in Chap. 4, the Schrödinger equation is exact only as a *global* equation. Its restriction to matter would require the global foliation of a classical spacetime. While this could still be chosen to proceed just locally (thus defining a 'finger of time'), the Wheeler–DeWitt equation describes entangled dynamics for global quantum states of matter *and* three-geometry.

How can the traditional concept of time (either in the form of many-fingered time, or as a parameter for the dynamics of global states) be recovered from the Wheeler–DeWitt equation? This requires concepts and methods discussed in Sect. 4.3, where quasi-classical quantities were shown to emerge dynamically and irreversibly by means of decoherence, but the dynamics has to be appropriately modified to suit the timeless Wheeler–DeWitt equation: classical time cannot emerge *in* classical time. Similarly, classical spacetime cannot have entered existence in a global quantum 'event' (which would have to presume time).

In the local field representation, the general Wheeler–DeWitt equation can be explicitly written in its gauge-dependent form (DeWitt 1967) as

$$-\frac{16\pi}{m_{\mathrm{P}}^2} \sum_{klk'l'} G_{klk'l'} \frac{\delta^2 \Psi}{\delta h_{kl} \delta h_{k'l'}} - \frac{m_{\mathrm{P}}^2}{16\pi} \sqrt{h}(R - 2\Lambda)\Psi + H_{\mathrm{matter}}\Psi = 0 \,, \quad (6.14)$$

when disregarding factor ordering. Here, Ψ is a functional of the *six* independent functions $h_{kl} = g_{kl}$ ($k, l = 1, 2, 3$), which represent the spatial metric on a hypersurface. The letter h (without indices) means their determinant, R their spatial Riemann curvature scalar. Λ is the cosmological constant, while $m_{\mathrm{P}} := 1/\sqrt{G}$ is the Planck mass. The hamiltonian density, H_{matter}, also depends on the metric by means of its kinetic energy terms. The matrix $G_{klk'l'} := (h_{kk'}h_{ll'} + h_{kl'}h_{lk'} - 2h_{kl}h_{k'l'})/2\sqrt{h}$ with respect to the six symmetric *pairs* of indices kl is DeWitt's 'superspace metric'. It has the *locally* hyperbolic signature $-+++++$.

The Wheeler–DeWitt equation (6.14) can assume this local form only because of its gauge degrees of freedom. Their elimination requires the wave functional to obey the three momentum constraints (see Sect. 5.4), in their quantum mechanical form written as $H_k \Psi = (\delta \Psi / \delta h_{kl})|_l = 0$, where $|_l$ is the covariant derivative with respect to the spatial metric h_{kl}. They represent *three* functional differential equations for the functional $\Psi[h_{kl}]$, which depends on six variables h_{kl} at each point. Integration of the Wheeler–DeWitt equation (6.14) under the constraints would then leave *two* functions as 'integration constants'. These two degrees of freedom at each space point may be regarded as representing the two physical components (polarizations) of the gravitational field. The momentum constraints are analogous to Gauß's law in electrodynamics, which also forms a constraint on the initial data when written in terms of the potential $\boldsymbol{A}$. However, the momentum constraints are 'secondary': they can only in special situations be solved analytically. In the superspace region that describes Friedmann-type universes, all but one of the infinity of negative kinetic energy terms of the Wheeler–DeWitt equation represent gauge degrees of freedom (see footnote 4 of Chap. 5). The remaining physical dynamics is *globally* hyperbolic.

I shall now assume that a solution of these coupled equations exists. Since the quantity $m_P^2/32\pi$ that appears in (6.14) as a formal dynamical mass is very large compared to dynamical masses contained in H_{matter}, one may conveniently analyze the solution by means of a Born–Oppenheimer approximation in analogy to molecular physics (Banks 1985). The matter wave function will then adiabatically depend on the massive gravity variables – even though there is no time-dependence. This situation resembles *small* molecules, which are usually found in their energy eigenstates (giving rise to rotational and vibrational bands) rather than in states representing quasi-classical motion. However, as a *novel* aspect of quantum cosmology, the matter degrees of freedom must now also describe observers, while molecules or other microscopic systems are observed from outside. Because of the adiabatic correlation of the observer with the quantum state of geometrodynamics (and that of other macroscopic variables), this quasi-classical state appears 'given' to him (see Sect. 4.6).

In order to obtain a semiclassical dynamical description of spacetime geometry, one may use the *ansatz*

$$\Psi[h_{kl}, x] = \exp\left[iS_0(h_{kl})\right]\chi(h_{kl}, x) , \qquad (6.15)$$

where x represents all matter degrees of freedom. S_0 is defined as a solution of the Hamilton–Jacobi equation of geometrodynamics (Peres 1962), with a self-consistent source term that is given as the expectation value $\langle\chi|T_{\mu\nu}|\chi\rangle$ of the matter states χ that are to be calculated along classical trajectories described by S_0. This is analogous to the description of *large* molecules, where the heavy atomic nuclei or ions are dynamically described by time-dependent classical orbits resulting from an effective potential that arises from an expectation value for electron wave functions, which in turn depend adiabatically on the nuclear positions. If the global boundary conditions are appropriate to justify the WKB approximation, the matter wave function χ may indeed depend adiabatically on the massive variables h_{kl} in the relevant regions of configuration space. As this spatial metric describes the three-geometry as the carrier of information about time along every WKB trajectory (similar to geometric optics), *the dependence of χ on h_{kl} may be regarded as a generalized physical time dependence* of the matter states.

Since the classical Hamilton–Jacobi equations describe an ensemble of dynamically independent trajectories in the configuration space of the three-geometries, the remaining equations for the matter states χ can be integrated along them. This procedure becomes particularly convenient after the exponential $\exp(iS_0)$ has been raised to the usual second order WKB approximation that includes a 'prefactor' which warrants the conservation of probability. In its local form (6.14), the Wheeler–DeWitt equation is reduced by the *ansatz* (6.15) to a *Tomonaga–Schwinger equation* (Lapchinsky and Rubakov 1979, Banks 1985):

$$i \sum_{klk'l'} G_{klk'l'} \frac{\delta S_0}{\delta h_{kl}} \frac{\delta \chi}{\delta h_{k'l'}} = H_{\mathrm{matter}} \chi . \tag{6.16}$$

Its LHS is (the local component of) a derivative of χ in the direction of the gradient ∇S_0 in the configuration space of three-geometries. Written as $i\nabla S_0 \cdot \nabla \chi =: i d\chi/d\tau$, it defines a many-fingered time parameter τ *for all trajectories* (that is, for all different spacetimes in the ensemble described by the specific solution S_0). Because of its dependence on the WKB approximation, τ may be called a 'WKB time'. Since the WKB wave function in general describes an extended superposition, its individual quasi-trajectories (corresponding to narrow wave tubes), correlated to their specific matter states χ, can only represent Everett *branches* of a general quantum universe. They are indeed decohered from one another by the adiabatic dependence on gravity of the uncontrollable microscopic degrees of freedom contained in χ (Sect. 4.3.5).

Equation (6.16) thus represents an effective *time-dependent* Schrödinger equation for matter. Higher orders of the WKB approximation lead to corrections to this Schrödinger dynamics (Kiefer and Singh 1991). The local fingers of time – represented by this local form – may be combined by integrating (6.16) over three-space (Giulini 1995). This integration elevates the local inner product $\nabla S_0 \cdot \nabla \chi$, that is, a sum over $k'l'$ by means of the Wheeler–DeWitt metric, to a global one. In this way it defines the progression of a reparametrizable *common global* dynamical time, valid for *all* spacetimes which are described by the Hamilton–Jacobi function S_0. Since τ is defined for all WKB trajectories (thus defining 'simultaneous' three-geometries on them), it also defines a *foliation of superspace* (Giulini and Kiefer 1994). Only the definition of spacetime coordinates requires lapse and shift functions N and N_k (Sect. 5.4), which then also define 'velocities' $\dot{h}_{kl}$ with respect to coordinate time according to

$$\dot{h}_{kl} = -N G_{klk'l'} \frac{\delta S_0}{\delta h_{k'l'}} + N_{k|l} + N_{l|k} . \tag{6.17}$$

The *complex ansatz* (6.15) is obviously essential for the result (6.16). The correct Wheeler–DeWitt wave function will in general have to be approximated by a superposition of *several* such WKB components. There is no reason to expect a complex solution for the complete Wheeler–DeWitt wave function that describes the Quantum Universe. So one may, in particular, have to replace (6.15) by its real part, $\Psi \to \Psi + \Psi^*$. The two terms may then *separately* obey (6.16) and its complex conjugate to an excellent approximation, similar to the various WKB trajectories (or wave tubes) which have to be superposed to form the extended wave front of geometric optics described by an appropriate Hamilton–Jacobi solution S_0 (or some higher order approximation S_1). This dynamical separation can again be understood as a consequence of decoherence by the microscopic matter degrees of freedom (Halliwell 1989, Kiefer 1992). It is interesting, as it represents a simple but fundamental example of *gauge symmetry breaking by Everett branching*. Note, however, that in contrast to the complex classical field in Sect. 6.1, the complex form of (6.16) characterizes superpositions in Hilbert space – not in the configuration space of fields.

The Born–Oppenheimer approximation with respect to the Planck mass is not always the most appropriate one. Macroscopic matter variables may be described by a WKB approximation, too, while certain geometric modes (graviton states) may not. These variables could then be shifted between S_0 and χ (see Vilenkin 1989). For example, Halliwell and Hawking (1985) applied the WKB approximation only to the monopoles α and Φ which characterize the quantum Friedmann universe (6.5). Their WKB solution would describe an ensemble of trajectories $a(t)$, $\Phi(t)$ in mini-superspace (Sect. 5.4). Wave functions for the (nonlocal) higher multipole amplitudes of matter and geometry can then again be obtained by means of a Tomonaga–Schwinger equation, similar to (6.16). This choice of *nonlocal* variables offers the advantage of separating physical degrees of freedom (in this case the pure tensor modes) from the remaining pure gauge modes. As mentioned in Sect. 5.4, the physical tensor modes all appear in the kinetic energy with a dynamical mass of the same sign as the matter modes, while only the monopole variable a (or $\alpha = \ln a$) has negative dynamical mass. In this model, the monopoles a and Φ are very efficiently decohered by the tensor modes (Zeh 1986b, Kiefer 1987, Barvinsky et al. 1999). Decoherence of local 'fluctuations' may lead to the formation of large scale structures of the Universe (Kiefer, Polarski and Starobinsky 1998).

The Tomonaga–Schwinger equation (6.16) justifies a dynamical time parameter on the basis of the timeless Wheeler–DeWitt equation, but not yet an arrow of time. Its solutions require 'initial' conditions with respect to the global time parameter τ. In order to describe our observed time-asymmetric Universe, these initial conditions must be responsible for the arrow(s) of time (in particular 'quantum causality', that has already been used when referring to decoherence). However, they cannot be freely postulated any more, but must be obtained from the complete Wheeler–DeWitt wave function, that depends on its own boundary conditions. For them, the hyperbolic nature of the Wheeler–DeWitt equation is essential. As discussed in Sect. 6.2.1, initial conditions may be posed for it at any fixed value of a time-like variable in superspace, such as α. While the initial condition for χ has to characterize an early Everett branch, the total wave function, which gives rise to the time arrow of this branching, must depend on a general boundary condition (perhaps just normalizability). 'Before' anything may evolve in classical time – as assumed when applying (6.16) in the forward direction, classical time must itself emerge from an appropriate initial condition that is a consequence of boundary conditions for the solution of (6.14). For example, if the dynamics in the form (6.6) is generalized by means of higher multipoles as

$$H = \frac{\partial^2}{\partial \alpha^2} + H_{\text{red}}^2 = \frac{\partial^2}{\partial \alpha^2} + \sum_i \left[-\frac{\partial^2}{\partial x_i^2} + V_i(\alpha, x_i) \right] + V_{\text{int}}(\alpha, \{x_i\}) , \quad (6.18)$$

where the x_i are now *all other* variables, the potential V_{int} may be highly asymmetric under a reversal of the sign of α. In particular, it could have a simple structure for $\alpha \to -\infty$, as indicated in (6.5), which might give rise to a 'simple' initial condition (SIC) in α (see Conradi and Zeh 1991). This may even explain the symmetric initial vacuum of Sect. 6.1. In the absence of any theory that describes the big bang singularity, we can only *assume* an appropriate simple structure of the total wave function in this limit (just as discussed for $t \to 0$ at the end of Chap. 4).

Inasmuch as the Tomonaga–Schwinger equation along a WKB trajectory in superspace describes *measurements*, it is practically useless for calculating 'backwards' in global time τ. One would have to know all (observed and unobserved) final *branches* of the total matter wave function χ (such as those that have unitarily arisen during measurements in the past). In the 'forward' direction, this global Schrödinger equation for matter has to be replaced by a master equation if decoherence of the quantum state of matter by gravity is relevant (see Sect. 4.3.4). This may explain the oft-proposed gravity-induced collapse as an apparent one in the 'usual' manner.

This limited applicability of a time-dependent Schrödinger or master equation that is based on WKB time would become particularly important for a recontracting universe (Sect. 5.3). If a master equation can be derived for all or most WKB trajectories with respect to that direction of τ that represents increasing a, it cannot remain valid along a classical spacetime history that leads to recontraction – cf. Fig. 6.3. Page (1985) and Hawking (1985) understandably arrived at the opposite conclusion when they described a recontracting universe by using WKB trajectories beyond the turning point (see the discussion following Zeh 1994). They thereby interpreted their semi-classical Feynman paths as representing an *ensemble of possible* cosmic histories that they justified by 'initial quantum uncertainties'. Their further treatment then neglects the final condition in τ that would be part of the initial condition in α, and give rise to formal recoherence along the trajectory even if quasi-trajectories of geometry *were* defined beyond the turning point.

Quantum cosmology requires a *consistent realistic interpretation* of quantum theory (Everett's, for example). It is often uncritically applied by using some pragmatic (for this purpose insufficient) interpretation, including a traditional concept of time. Let me therefore briefly mention consequences of a timeless wave function on Bohm's quantum mechanics, which were first discussed for different reasons by Julian Barbour (1994a,c):

In addition to a time-dependent Everett wave function, Bohm's theory postulates the existence of an ensemble of trajectories in a classical configuration space that describes particles and fields (see Sect. 4.6). If quantum gravity is taken into account, this configuration space must include three-geometries. In a time-less theory, the trajectories degenerate into an ensemble of fixed states (points), assumed to possess 'statistical weights' according to $|\Psi|^2$. In contrast to Bohm's time-dependent theory, this is *no longer an initial condition* that would have to be preserved by the presumed unobservable dynamics for the Bohm trajectories.

The structure of the Wheeler–DeWitt wave function in the range of applicability of a WKB approximation then statistically favors those classical states which lie on *apparent trajectories*. This result is very similar to Mott's (1929) description of α-particle tracks in a cloud chamber, where the 'stationary' (static) wave function suppresses configurations that describe droplets not approximately lying along particle tracks. Barbour calls these preferred states 'time capsules', since they represent consistent memories (without corresponding histories). In Barbour's words: "time is in the instant" (in the state) "– the instant is not in time" (in a history). If all classical states in the ensemble are regarded as 'real' (precisely as all past and future states are assumed to form a real one-dimensional history in the conventional block universe description), they now form a multi-dimensional rather than a one-dimensional continuum. One may even say that time is *replaced* by the wave function in this picture.

In contrast to the Everett interpretation, Bohm's theory presumes these classical configurations as part of fundamental reality, which must include observers. Each electron in a molecule, for example, is then assumed to possess a definite position in every actual state (though not any velocity or momentum). Since this particle position is *not* part of a memorized or documented (real or apparent) history according to this interpretation, we are only led to *believe* that it 'actually exists' as a wave function. The intrinsic dynamics of the static Wheeler-DeWitt wave function has the consequence that the electron's effects on measurement devices are dynamically 'caused' by *all* its positions in the support of the wave function (in dependence of the latter's amplitude) – not by a one-dimensional history. This picture would explain why the arena for the wave function is a classical configuration space, although most problems and disadvantages of Bohm's theory (see Zeh 1999b) persist, and even new ones may arise. Why should there be arbitrary global simultaneities representing actual elements of reality, while 'actuality' seems to be meaningful only with respect to local observers?

General Literature: Anderson 2006, Kiefer 2007.

6.2.3 Black Holes in Quantum Cosmology

During the early days of general relativity, the spacetime region behind a black hole horizon was regarded as meaningless, since it is inaccessible to observers in the external region. From their positivistic point of view, it would 'not exist'. Later one realized that world lines, including those of observers, can be smoothly continued beyond the horizon, where they would hit the singularity within finite proper time. The new conclusion, that the internal

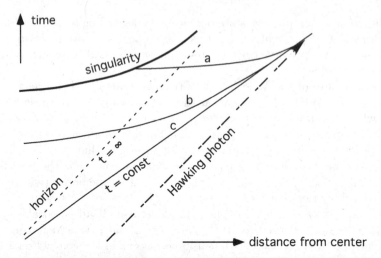

Fig. 6.4. Various kinds of simultaneities for a spherical black hole in a Kruskal type diagram: (a) hitting the singularity, (b) entering only the regular internal region, (c) completely remaining outside (Schwarzschild coordinate t). *Any* Schwarzschild time, for example $t = t_{turn}$, may be identified with $t = 0$ (a horizontal line in the diagram) regardless of the time of the observed collapse. No horizon forms on the Schwarzschild simultaneities, which are complete for the external universe. (From Zeh 2005c)

regions of black holes are physically 'regular' except at the singularity (hence for limited time only), seems to apply as well to Bekenstein–Hawking black holes until they disappear (see Sect. 5.1). However, arguments indicating a genuine (possibly dramatic) quantum nature of the event horizon have also been raised ('t Hooft 1990, Keski-Vakkuri et al. 1995, Li and Gott 1998).

While a consistent quantum description of black holes has not as yet been presented, attempts were mostly based on semiclassical methods. (For an overview see Kiefer 2007.) When combined with quantum cosmology, they may lead to important novel consequences, which seem to revive the early doubts in the meaning and existence of black hole interiors.

Consider the Schwarzschild metric (Fig. 5.1) as far as it is relevant for a black hole formed by collapsing matter, such that the Kruskal regions III and IV do not occur (Fig. 5.3a). Its dynamical (3+1) description in terms of three-geometries depends in an essential way on the choice of a foliation (see Fig. 6.4, or the Oppenheimer–Snyder model described in Box 32.1 of Misner, Thorne and Wheeler 1973). Three-geometries which intersect the event horizon may spatially extend to the singularity at $r = 0$, and thus render the global quantum states that they carry prone to dynamical indeterminism or consequences of a future theory that may avoid singularities. In contrast, a foliation according to Schwarzschild time t would describe regular three-geometries for $t < \infty$, which could be continued *in time* beyond $t = \infty$ by means of the new

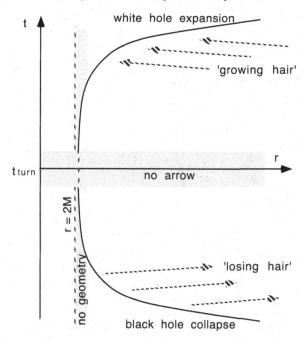

Fig. 6.5. Classical trajectory of a collapsing dust shell or the surface of a collapsing star (*solid curve*) in a thermodynamically symmetric recontracting universe. It is represented here in compressed Schwarzschild coordinates as used in Fig. 5.1, with the Schwarzschild metric now being valid only to the right of the star's surface. Because of the scale compression, light rays appear almost horizontal in the diagram. For $t > t_{\text{turn}}$, advanced radiation from the formal past would focus onto the black hole, which must now re-expand and grow hair in this scenario, while observers would experience time in the opposite direction. No horizon ever forms: the region $r < 2M$ (which is later than $t = \infty$) would arise only if gravitational collapse continued forever in a classical manner. Because of the drastic quantum effects close to the turning point of a Friedmann universe (see Fig. 6.3), there are in general only 'probabilistic' connections between quasi-classical branches in the expansion and contraction eras of the Universe. (From Kiefer and Zeh 1995)

time coordinate r (with physical time growing with decreasing r for $r < 2M$). According to this foliation, the black hole interior with its singularity would always remain in our formal future, and the singularity must be irrelevant for Hawking radiation. In the pair creation picture, the negative-energy partner is absorbed to the spacetime region close to what appears as a horizon until this is completely transformed into radiation. Therefore, this foliation seems to be appropriate for the formulation of a cosmological boundary condition (in superspace), that may explain the master arrow of time.

For further discussion now assume that the expansion of the classical universe on which this diagram is based is reversed at a finite Schwarzschild time

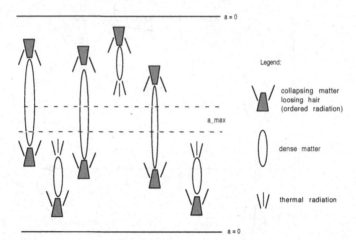

Fig. 6.6. Quasi-classical picture (using Schwarzschild coordinates) of a thermodynamically T-symmetric quantum universe which contains black holes, white holes, and black-and-white holes that re-expand by anti-causal effects. Instead of horizons and singularities, there are merely spacetime regions of large curvature ('dense matter') in this scenario. Because of their strong time dilation, they may serve as a short cut in proper time between big bang and big crunch (or between the presumed eras of opposite arrows of time). 'Information-gaining systems' could not thereby survive as such. In quantum cosmology there is no unique connection between quasi-classical histories (Everett branches) represented by the upper and lower halves of the figure, but there is no need for a violation of conservation laws

$t = t_{\mathrm{turn}}$ that is much larger than the time of the effective gravitational collapse (losing hair – see Fig. 6.5). No horizon yet exists on the Schwarzschild simultaneity $t = t_{\mathrm{turn}} < \infty$. If the cosmic time arrow does change direction (while the quasi-classical universe passes through an era of thermodynamical indefiniteness), the gravitationally collapsing matter close to the expected horizon will very soon (in terms of its own proper time) enter the era where radiation is advanced in the sense of Chap. 2. The black hole can then no longer 'lose hair' by emitting retarded radiation; it must instead 'grow hair' in an anti-causal manner (Fig. 6.6). According to a 'time-reversed no-hair theorem' it has to re-expand when the Universe starts recontracting (Zeh 1994, Kiefer and Zeh 1995).

This scenario does not contradict the geometrodynamical theorems about a monotonic growth of black hole areas, since no horizons ever form. A classical spacetime will not even exist close to the 'turning of the tide'. Here, decoherence is competing with recoherence before being replaced by it. Only region I of Fig. 5.1 is then realized. Events which appear 'later' than t_{turn} in the classical picture are 'earlier' in the sense of the intrinsic dynamics of the Wheeler–DeWitt equation (6.6) – and therefore also in the thermodynamical sense if this is based on an intrinsic initial condition. This quantum cosmological model describes an *apparent* (quasi-classical) two-time Weyl tensor or similar condition (see Fig. 6.6). In quantized general relativity, the two apparently different boundaries are identical, and thus represent *one and the same* boundary condition. The problem of their consistency (Sect. 5.4) is reduced to the intrinsic 'final' condition of normalizability for $a \to \infty$.

The description used so far in this section does not apply directly to a forever-expanding universe, where the arrow would preserve its direction along a complete quasi-trajectory from $a = 0$ to $a = +\infty$. The Wheeler–DeWitt wave function is then *not* normalizable for $a \to \infty$. However, one may require this wave function to vanish on *all* somewhere-singular three-geometries by a symmetric generalization of the Weyl tensor hypothesis. Such a condition has been confirmed to apply to a simple quantum model of a collapsing thin spherical matter shell (Hájíček and Kiefer 2001). In more realistic cases it would again lead to important thermodynamical and quantum effects close to event horizons (Zeh 1983), and drastically affect (or even exclude) the possibility of continuing a quasi-classical spacetime beyond them. These consequences would be unobservable in practice by external observers, since the immediate vicinity of a future horizon remains outside their backward light cones for all finite future. In order to receive information from the vicinity of a future horizon, one has to come dangerously close to it, and thus participate in the extreme time dilatation (see Fig. 5.2, where the light cone structure is made evident, while distances are strongly distorted).

These conclusions seem again to throw serious doubts on the validity of a classical continuation of spacetime into black hole interiors (see also Kiefer 2004 or Zeh 2005a). Event horizons in classical general relativity may signal the presence of drastic thermodynamical and quantum effects rather than representing 'physically normal' regions of spacetime. While their observable consequences depend on the world lines of detectors or observers (their acceleration, in particular), global quantum states, such as a specific 'vacuum', are invariantly defined – though not invariantly observed (Sect. 5.2). These global states may define an objective arrow of time, including 'quantum causality' (responsible for decoherence), by means of a fundamental boundary condition for the Wheeler–DeWitt wave function.

Epilog

Four weeks before his death, Albert Einstein wrote in a letter of condolence to the family of his life-long friend Michael Besso (Dukas and Hoffman 1979):[5] "For us believing physicists, the division into past, present and future has merely the meaning of an albeit obstinate illusion." There is no doubt that Einstein meant this remark seriously. Evidently, it refers to the four-dimensional ('static') spacetime picture of a 'block universe' that his theory of relativity uses so efficiently. This picture seems to be at variance with the experience of a *present passing through time* (the 'flow' or 'passage of time'). In contrast, the relativistic spacetime framework contains only a concept of *local events* (points in spacetime), which may be regarded as a continuum of dynamically related *here-and-nows*. Because of these dynamical relations, characterized by time-symmetric local laws, a *local present* can be viewed as 'moving' along the world line of an observer. His personal history is a succession of strongly correlated (local clusters of) events, dynamically controlled by proper time, while a global dynamical state would depend on an arbitrary foliation of the spacetime that is characterized by its invariant metric structure.

In Hermann Weyl's words: "The objective world simply *is*; it does not happen. Only to the gaze of my *consciousness, crawling upward* along the life line of my body, does a section of this world come to life as a fleeting image in space that continuously changes in time" (my italics). Any objective (classical) description of the locally experienced world must therefore treat space and time on an equal footing: it does not contain the concept of an *objective global present*. For this reason, Huw Price (1996) chose the subtitle 'A View from Nowhen' for his book on time's arrow. However, whether the objective world 'simply is', or rather 'comes into being', seems nonetheless to be a pure matter of words. Weyl's 'is' (the block universe picture) does not exclude a dynamical evolution (see Sect. 5.4).

---

[5] *"Für uns gläubige Physiker hat die Scheidung zwischen Vergangenheit, Gegenwart und Zukunft nur die Bedeutung einer wenn auch hartnäckigen Illusion."*

The four- (or higher-) dimensional 'static' *view* is by no means specific to the theory of relativity. It does not even require deterministic laws, as it was already used by St. Augustine in his *Confessiones*. He regarded it as a divine world view – presumably since he understood it as including all details – not merely as a conceptual framework. A mortal physicist may at least *conceive of* the future history of the world (though with less confidence in the details than in those of the past). Even Laplace could not have expected his model world to be determinable *in practice* (see Sect. 3.3); he had to assume an extra-physical demon of unlimited capacities for this purpose. The argument that even the *macroscopic* future cannot in general be known to physical systems, such as humans or computers, should not be confused with a conceivable indeterminism of the dynamical laws (as is often done in the theory of chaos – see the lucid article by Bricmont 1996).

The peculiarity of the subjective present, often mistaken as part of an objective 'structure of time', was emphasized by Einstein in a conversation with Carnap. According to Carnap (1963), "Einstein said that the problem of the Now worried him seriously. He explained that the experience of the Now means something special for man, something essentially different from the past and the future, but that this important difference does not and cannot occur within physics. That this experience cannot be grasped by science seemed to him a matter of painful but inevitable resignation." So he concluded "that there is something essential about the Now which is just outside the realm of science."

Carnap emphasized, however, that Einstein agreed with him (in contrast to Bergson and other philosophers) that this situation does *not* indicate a *defect of the physical concept of time*. (The non-confirmation of a prejudice is easily viewed as a defect!) The situation should rather be understood as reflecting the undefined role of the observer, that must characterize the fundamental and underivable here-and-now of *subjective* reality in the form of a local psycho-physical parallelism, for example. Objective reality (the 'divine world picture' of a block universe) must instead always remain a hypothesis – but this successful ('heuristic') fiction[6] describes an empirically founded *asymmetry* in time that does *not* require the concept of a present which flows in time.

---

[6] This notion of a 'fiction' is quite compatible with that of reality, even though it emphasizes the impossibility of *proving* the existence of a real world. For example, Einstein's *gläubiger Physiker* (believing physicist) may express the *belief* in an objective reality (in Einstein's case preferentially a local one) to be described by the physical formalism. As an admirer of Hume (but not of Kant – see Franck 1949) Einstein was clearly aware of this hypothetical character of reality. The evidence that Nature appears comprehensible was regarded by Einstein as "the most incomprehensible thing about Nature". One should here recall that Descartes and Hume raised their doubts and criticism, which essentially forces us into fictionalism, against attempts to obtain absolute *certainty* in the empirical sciences – not against any *conceivable* reality – see also d'Espagnat (1995).

The role of time is very different in quantum theory because of the latter's probabilistic appearance. The 'openness of the future' has even been regarded as its most fundamental novel aspect (von Weizsäcker 1982), although its origin and meaning remain controversial. Heisenberg spoke of the trajectory of a particle 'coming into being by human observations' (see the end of the Introduction), while the later Niels Bohr seems to have regarded *objective* quantum events as occurring 'out of the blue' in the measurement device (see Ulfbeck and Bohr 2001). Similar ideas about a fundamental concept of *becoming* in Nature were upheld for somewhat different (though also questionable) reasons by Prigogine (1980).

It is evident that these various opinions are based on different interpretations of quantum theory (see Sect. 4.6). For example, the wave function has been regarded as representing mere 'potentiality' or some novel fundamental concept of information ('it from bit' – see Wheeler 1994). These concepts would then asymmetrically apply only to the future, while the past is presumed to be given. Some of them may in fact represent no more than words (see Tegmark 1998). In particular, if quantum theory were just a stochastic theory, the block universe picture would still apply to the 'divine world view', as mentioned above, while quantum theory is in conflict with *any* local reality.

If the wave function (or some more general concept of superpositions) is the correct kinematical concept of quantum theory (complete, in particular, to define entropy as a measure of irreversibility), its dynamics must be essential and sufficient to analyze the objective arrow of time – regardless of any further interpretation. This dynamics is deterministic as far as the Schrödinger equation holds, but probabilistic whenever a collapse of the wave function has to be taken into account. Although the collapse probabilities have been claimed to be time-symmetric (Aharonov, Bergmann and Lebowitz 1964), since $|\langle a|b\rangle| = |\langle b|a\rangle|$ for any two states a and b, the structure of initial and final states of a probabilistic event is usually very different – see (4.56) and footnote 2 of Chap. 5. A generic collapse would *reduce* the entanglement between subsystems, as it projects onto definite 'pointer states', while it is usually preceded by a much larger *increase* of entanglement (decoherence). This latter asymmetry can be explained by means of an appropriate initial condition for the universal wave function (the absence of initial entanglement). In the case of a collapse, symmetry could be restored only if this collapse were allowed also to 'create' nonlocal entanglement in an acausal manner in order to reverse (4.56).

Could the subjective experience of a present that seems to flow in turn induce an *apparent* time asymmetry that had no counterpart in the real world? Einstein's above-quoted remark is often interpreted as supporting this idea of an arrow of time as an illusion. It is, therefore, often mentioned by the fundamental information-theoretical school of statistical mechanics (Sect. 3.3.1), or in favor of an *extra-physical* concept of (growing) 'human knowledge' – corresponding to Heisenberg's 'idealistic' interpretation of the wave function.

I have tried to explain in various chapters of the book that the concept of entropy is in fact observer-related by means of a *relevance concept*. This observer-relatedness of the macroscopic description (which includes 'pointer states' of measurement devices) is particularly important in quantum theory, and – as it turns out – even for the emergence of a classical concept of time from a timeless quantum world. However, the observed time-asymmetry could always be traced back (at least in principle) to the asymmetric structure of an objective physical reality, that according to present knowledge is represented by a nonlocal quantum world.

Memories, in particular, have to be stored in physical form, and are then correlated with sources in their past (they are 'retarded'). This drastic asymmetry may be sufficient to explain the *apparent* flow of time once there is a psycho-physical parallelism based on a presumed local *moment of awareness*. Only this (*not* necessarily asymmetric) concept of a local present is fundamentally subjective, while the asymmetry between past and future directions is part of objective reality. What we usually call the preserved *identity* of a person (who changes considerably during his lifetime) is 'in reality' nothing but a particularly strong and robust 'causal' correlation between *different* local physical states which represent the individual carriers of a subjective present. As pointed out by Einstein and Carnap, it is the *here-and-now* subjectivity as the center of all awareness that goes beyond objective reality, while it must severely affect our perception of the 'real world'.

The essential novel aspect of quantum theory is its nonlocality. The dislocalization of superpositions with increasing time forces certain causal chains, which may also represent observers, to exist only in dynamically autonomous *components* of the global quantum state. Since these components then branch in our asymmetric quantum world, the quasi-classical world *appears* indeterministic to these branching observers. Entanglement entropy of 'systems' may appear to be an objective quantity that is defined by the global wave function, but locality (essential to characterize systems) already represents a non-trivial relevance concept that can be justified only by the locality of the observer, which is *facilitated* by the locality of dynamics.

Appendix: A Simple Numerical Toy Model

Various elementary models which are intended to illustrate statistical methods in physics can be found in the literature. Well known are the *urn model* (Ehrenfest and Ehrenfest 1911) and the *ring model* (Kac 1959). Both are based on stochastic dynamical assumptions, applied in a given time direction. The model to be discussed below is deterministic, and based on special initial conditions. Its main purpose is to illustrate the concept of a Zwanzig projection by means of the simple example of spatial coarse-graining. This coarse-graining is used for the definition of entropy, and, in a second step, also to construct a master equation. However, it then fails to represent a good approximation, thus demonstrating the importance of *dynamical* properties of a Zwanzig projection if it is to be useful.

The model is defined by a swarm of n_P free particles ($i = 1, \ldots, n_P$), moving with fixed velocities $v_i = v_0 + \Delta v_i$ on a 'ring' (a periodic interval) divided into n_S 'unit cells' ($j = 1, \ldots, n_S$), where the Δv_i's form a homogeneous random distribution in the range $\pm \Delta v/2$. Coarse-graining is defined by averaging over cells. All particles are assumed to start from random positions in the cell $j = 1$.

The Zwanzig concept of relevance is furthermore assumed not to distinguish between the (distinguishable) particles, thus establishing an 'occupation number representation on unit cells'. Since momenta are conserved here, only the *spatial* distribution of particles has to be considered. If n_j particles are in the jth cell, the entropy according to this Zwanzig projection is given by

$$S = -\sum_{j=1}^{n_S} \frac{n_j}{n_P} \ln \frac{n_j}{n_P} \ . \tag{6.19}$$

It vanishes in the initial state, while its maximum, $S^{\max} = \ln n_S$, holds for equipartition, $n_j = n_P/n_S$. With respect to this coarse-grained entropy, the model has a statistical Poincaré recurrence time $t_{\text{Poincaré}} = n_S^{n_P - 1}$, while the relaxation time, required for the approach to equipartition, is only of order $n_S/\Delta v$. It is so short, since this coarse-graining is not in any way robust. Other equilibrium distributions could be constructed by using intervals of different lengths.

The first plot of the *Mathematica* notebook[7] below (plot1) shows the entropy evolution for $n_P = 100$ and $n_S = 20$ during the first 2000 units of time. The relaxation time scale is of the order 1000 units. Only integer times are plotted in order to eliminate otherwise disturbing lattice effects of the model. At some later time (see plot2), the coarse-grained representation no longer reveals any information about the existence of a low-entropy state in the recent past, although this information must still exist, since motion can be reversed in this deterministic model.

One can now simply enforce a two-time boundary condition (Sect. 5.3.3) by restricting all relative velocities Δv_i to integer multiples of n_S divided by a large integer (e.g., by rounding them off at a certain figure). Consequences of this change are negligible for small and intermediate times, although the evolution is now exactly periodic (on an interval of 200 000 units in the chosen numerical example, which is a very small fraction of the statistical recurrence time). *Relative* entropy minima may occur at simple rational fractions of this interval (such as at 2/5 of 200 000 in plot3).

When using the coarse-graining dynamically, one obtains a master equation for *mean* occupation numbers $\bar{n}_j(t)$ in an ensemble of individual solutions. For $v_0 = 0$ it would read [see (4.45)]

$$\frac{d\bar{n}_j}{dt} = \lambda(\bar{n}_{j-1} + \bar{n}_{j+1} - 2\bar{n}_j) , \tag{6.20}$$

where λ (in this case given by $\Delta v/8$) is the mean rate for particles to move by one unit in either direction. For $v_0 \neq 0$, the result would hold in the center-of-mass frame. The lower smooth curve of plot4 shows the resulting monotonic increase of ensemble entropy, compared with the individual (fluctuating) solution of plot1.

Evidently, the two curves agree only for about the first $2/\Delta v$ units of time. This demonstrates that this Zwanzig projection is not very appropriate for *dynamical* purposes, since some fine-grained (neglected) information remains dynamically relevant: the *individually* conserved velocities lead here to relevant correlations between position and velocity. The master equation, which does not distinguish between individual particles, allows them even to change their direction of motion. A slightly improved long-time approximation can therefore be obtained by assuming all particles to diffuse in only one direction (upper smooth curve of plot4) – equivalent to using a reference frame that moves with a velocity $-\Delta v/2$ in the center-of-mass frame. It still neglects dynamically relevant correlations resulting from the fact that the most distant particles continue travelling fastest. These correlations remain dynamically relevant during relaxation, that is, until most particles have diffused once around the ring. This can be recognized in plot4.

The master equation (6.20) may also represent an ensemble of individually stochastic (indeterministic) histories $n_j(t)$, described by a Langevin equation.

[7] Programming aid by Erich Joos is acknowledged.

Notebook: A Toy Model

1. Definitions

```
nS = 20;
nP = 100;
v0 = 1.;
Δv = .02;

shannon[x_] := If[x == 0, 0., N[x Log[x]]]

entropMax = Log[nS] // N

2.995732274

poincareTime = nS^(nP - 1) // N

6.338253001 × 10^128

relaxTime = nS / Δv

1000.
```

2. Initial Values

```
SeedRandom[2]

v = Table[v0 + Δv (Random[] - .5), {nP}];

x0 = Table[Random[], { nP}];

<< "Statistics`DescriptiveStatistics`"

{Mean[x0], Mean[v], Variance[v]}

{0.5156437577, 0.9997682434, 0.00003356193642}
```

3. Exact Model

```
entropy[t_] := Module[{cellnb, n, e},

  (cellnb = Ceiling[Mod[v t + x0, nS]];

  Do[n[i] = Count[cellnb, i], {i, nS}];

  e = N[Log[nP] - (Σ_{j=1}^{nS} shannon[n[j]]) / nP] ) ]
```

```
plot1 = ListPlot[Table[{t, entropy[t]}, {t, -50, 2000, 5}],
    PlotJoined → True, Frame → True, PlotRange → {0, entropMax}];
```

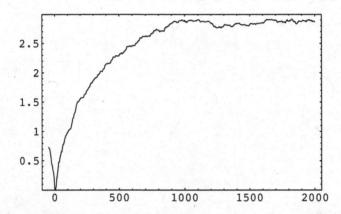

```
plot2 = ListPlot[Table[{t, entropy[t]}, {t, 10000, 12000, 5}],
    PlotJoined → True, Frame → True, PlotRange → {0, entropMax}];
```

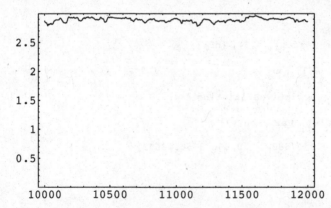

4. Two-Time Boundary Condition

$$v = \frac{\text{Floor}[10000\, v]}{10000};$$

```
period = 10000 nS
```

200000

```
plot3 = ListPlot[Table[{t, entropy[t]}, {t, 79000, 81000, 5}],
   PlotJoined → True, Frame → True, PlotRange → {0, entropMax}];
```

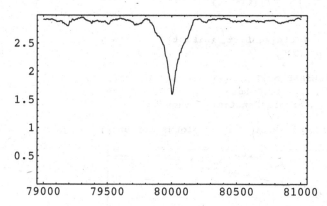

5. Master Equation

$$\lambda = \Delta v\, /\, 8$$

0.0025

```
equList = Table[n[j]'[t] == λ (-2 n[j][t]
      + n[Mod[j, nS] + 1][t] + n[Mod[j - 2, nS] + 1][t]), {j, nS}];
```

```
iniList = Join[ {n[1][0] == nP}, Table[n[j][0] == 0., {j, 2, nS}] ];
```

```
nSolution = NDSolve[Join[equList, iniList], Table[n[j],
    {j, nS}], {t, 0, 2000}];
```

$$\texttt{entropyMaster[t\_] := N[Log[nP]] - }\frac{\sum_{j=1}^{nS} \texttt{shannon[n[j][t]]}}{nP}$$

```
/. nSolution // First
```

```
plotMasterShort = ListPlot[Table[{x, entropyMaster[x] },
    {x, 0, 2000, 20}], PlotJoined -> True, PlotRange ->
    {0, entropMax}, DisplayFunction -> Identity];
```

```
equList = Table[n[j]'[t] == 4 λ (-n[j][t]
        + n[Mod[j - 2, nS] + 1][t]), {j, nS}];
```

```
nSolution = NDSolve[Join[equList, iniList], Table[n[j],
    {j, nS}], {t, 0, 2000}];
```

```
plotMasterLong = ListPlot[Table[{x, entropyMaster[x] },
    {x, 0, 2000, 20}], PlotJoined -> True, PlotRange ->
    {0, entropMax}, DisplayFunction -> Identity];
```

```
plot4 = Show[plot1, plotMasterShort, plotMasterLong];
```

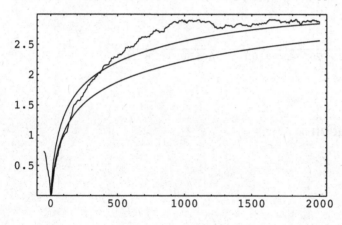

References

References marked with **(WZ)** can also be found (in English translation if applicable) in Wheeler and Zurek 1983, those with **(LR)** in Leff and Rex 1990. Some of my own (p)reprints are available through www.zeh-hd.de. The page of this book on which the reference is quoted is added in square brackets.

Aharonov, Y., Bergmann, P.G., and Lebowitz, J.L. (1964): Time symmetry in the quantum process of measurement. Phys. Rev. **134**, B1410 – **(WZ)** – [131,201]

Aharonov, Y., and Vaidman, L. (1991): Complete description of a quantum system at a given time. J. Phys. **A24**, 2315 – [126]

Albrecht, A., (1992): Investigating decoherence in a simple system. Phys. Rev. **D46**, 5504 – [133]

Albrecht, A. (1993): Following a "collapsing" wave function. Phys. Rev. **D48**, 3768 – [133]

Alicki, R., and Lendi, K. (1987): *Quantum Dynamical Semigroups and Applications* (Springer) – [118]

Anderson, E. (2006): Emergent semiclassical time in quantum gravity. E-print gr-qc/0611007 and gr-qc/0611008 – [193]

Anglin, J.R., and Zurek, W.H. (1996): A precision test of decoherence. E-print quant-ph/9611049 – [110]

Arndt, M., Nairz, O., Vos-Andreae, J., Keler, C., van der Zouw, G., and Zeilinger, A. (1999): Wave–particle duality of C_{60} molecules. Nature **401**, 680 – [104]

Arnol'd, V.I., and Avez, A. (1968): *Ergodic Problems of Classical Mechanics* (Benjamin) – [55]

Arnowitt, R., Deser, S., and Misner, C.W. (1962): The dynamics of general relativity. In: *Gravitation: An Introduction to Current Research*, ed. by Witten, L. (Wiley) – [161]

Ashtekhar, A. (1987): New Hamiltonian formulation of general relativity. Phys. Rev. **D36**, 1587 – [178]

Atkinson, D. (2006): Does quantum electrodynamics have an arrow of time? Stud. Hist. Phil. Mod. Phys. **37**, 528 – [3]

Baierlein, R.F., Sharp, D.H., and Wheeler, J.A. (1962): Three-dimensional geometry as carrier of information about time. Phys. Rev. **126**, 1864 – [163]

Balian, R. (1991): *From Microphysics to Macrophysics* (Springer) – [68]

Banks, T. (1985): TCP, quantum gravity, the cosmological constant, and all that .... Nucl. Physics **B249**, 332 – [186,189,190]

Barbour, J.B. (1986): Leibnizian time, Machian dynamics and quantum gravity. In: *Quantum Concepts in Space and Time*, ed. by Penrose, R., and Isham, C.J. (Clarendon Press) – [13,179]

Barbour, J.B. (1989): *Absolute or Relative Motion?* (Cambridge University Press) – [16]

Barbour, J.B. (1994b): The timelessness of quantum gravity II. The appearance of dynamics in static configurations. Class. Quantum Grav. **11**, 2875 – [192]

Barbour, J.B. (1999): *The End of Time* (Weidenfeld & Nicolson) – [13,16,165,170]

Barbour, J.B., and Bertotti, B. (1982): Mach's principle and the structure of dynamical theories. Proc. R. Soc. London **A382**, 295 – [165]

Barbour, J.B., and Pfister, H. (Eds.) (1995): *Mach's Principle: From Newton's Bucket to Quantum Gravity* (Birkhäuser, Basel) – [13,163]

Bardeen, J.M., Carter, B., and Hawking, S.W. (1973): The four laws of black hole mechanics. Comm. Math. Phys. **31**, 161 – [146]

Bardeen, J., Cooper, L.N., and Schrieffer, J.R. (1957): Theory of superconductivity. Phys. Rev. **108**, 1175 – [151]

Barrow, J.D., and Tipler, F.J. (1986): *The Anthropic Cosmological Principle* (Oxford University Press) – [84]

Bartnik, R., and Fodor, G. (1993): On the restricted validity of the thin sandwich conjecture. Phys. Rev. **D48**, 3596 – [162]

Barvinsky, A.O., Kamenshchik, A. Yu., Kiefer, C., and Mishakov, I.V. (1999): Decoherence in quantum cosmology at the onset of inflation. Nucl. Phys. **B551**, 374 – [187,191]

Bekenstein, J.D. (1973): Black holes and entropy. Phys. Rev. **D7**, 2333 – [146]

Bekenstein, J.D. (1980): Black-hole thermodynamics. Physics Today **33**, 24 – [148]

Belavkin, V.P. (1988): Nondemolition measurements, nonlinear filtering and dynamic programming of quantum stochastic processes. In: *Modeling and Control of Systems*, ed. by Blanquière, A. (Springer) – [118]

Bell, J.S. (1964): On the Einstein–Podolsky–Rosen paradox. Physics **1**, 195 – **(WZ)** – [96]

Bell, J.S. (1981): Quantum mechanics for cosmologists. In: *Quantum Gravity II*, ed. by Isham, C.J., Penrose, R., and Sciama, D.W. (Clarendon Press) – [127]

Beller, M. (1996): The conceptual and the anecdotal history of quantum mechanics. Found. Phys. **26**, 545 – [125]

Beller, M. (1999): *Quantum Dialogue* (University of Chicago Press) – [125]

Bennett, C.H. (1973): Logical reversibility of computation. IBM J. Res. Dev. **17**, 525 – **(LR)** – [74,75,76]

Bennett, C.H. (1987): Demons, engines, and the second law. Sci. Amer. **257**, 108 – [75,76]

Bennett, C.H., and Landauer, R. (1985): The fundamental physical limits of computation. Sci. Amer. **253**, 38 – [76]

Bertalanffi, L. von (1953): *Biophysik des Fließgleichgewichts* (Vieweg) – [77]

Birrell, N.D., and Davies, P.C.W. (1983): *Quantum Fields in Curved Space* (Cambridge University Press) – [153]

Bläsi, B., and Hardy, L. (1995): Reality and time symmetry in quantum mechanics. Phys. Lett. **A207**, 119 – [132]

Blyth, W.F., and Isham, C.J. (1975): Quantization of a Friedmann universe filled with a scalar field. Phys. Rev. **D11**, 768 – [181]

Böhm, A. (1978): *The Rigged Hilbert Spaces and Quantum Mechanics* (Springer) – [119]

Bohm, D. (1952): A suggested interpretation of the quantum theory in terms of "hidden" variables. Phys. Rev. **85**, 166; ibid. **85**, 180 – [126]

Bohm, D., and Bub, J. (1966): A proposed solution of the measurement problem in quantum mechanics by a hidden variable theory. Rev. Mod. Phys. **38**, 453 – [118]

Bohm, D., and Hiley, B. (1993): *The Undivided Universe: An Ontological Interpretation of Quantum Mechanics* (Routledge and Kegan) – [126]

Bohr, N. (1928): The quantum postulate and the recent development of atomic theory. Nature **121**, 580 – **(WZ)** – [114]

Bojowald, M. (2003): Initial Conditions for a Universe. Gen. Rel. Grav. **35**, 1877 – 182]

Boltzmann, L. (1866): Über die mechanische Bedeutung des zweiten Hauptsatzes der Wärmetheorie. Wiener Ber. **53**, 195 – [43]

Boltzmann, L. (1896): *Vorlesungen über Gastheorie* (Barth, Leipzig); Engl. transl. by Brush, S.G. (Univ. Cal. Press, 1964) – [43,155]

Boltzmann, L. (1897): Über irreversible Strahlungsvorgänge II. Berliner Ber., 1016 [19]

Bondi, H. (1961): *Cosmology* (Cambridge University Press) – [153]

Borel, E. (1924): *Le hasard* (Alcan, Paris) – [56]

Born, M. (1926): Das Adiabatenprinzip in der Quantenmechanik. Z. Physik **40**, 167 – [92]

Born, M. (1955): Continuity, determinism, and reality. Kgl. Danske Videnskab. Mat. Fys. Medd. **30**, 2 – [72]

Borzeskowski, H.-H. von, and Treder, H.-J. (1988): *The Meaning of Quantum Gravity* (Reidel) – [177]

Boulware, D.G. (1980): Radiation from a uniformly accelerated charge. Ann. Phys. (N.Y.) **124**, 169 – [31,34,150]

Bouwmeester, D., Ekert, A., and Zeilinger, A. (2000): *The Physics of Quantum Information* (Springer) – [108]

Brady, P.R., Moss, I.G., and Myers, R.C. (1998): Cosmic censorship: As strong as ever. Phys. Rev. Lett. **80**, 3432 – [143]

Bricmont, J. (1996): Science of chaos or chaos in science? In: *The Flight from Science and Reason*, ed. by Gross, P.R., Levitt, N., and Levis, M.W. (New York Acad. Sci., Vol. 755) – [72,200]

Brillouin, L. (1962): *Science and Information Theory* (Academic Press) – [72,75,160]

Bronstein, M., and Landau, L. (1933): Über den zweiten Wärmesatz und die Zusammenhangsverhältnisse der Welt im Großen. Sowj. Physik **4**, 114 – [82]

Brown, J.D., and Kuchař, K.V. (1995): Dust as a standard of space and time. Phys. Rev. **D51**, 5600 – [162]

Brun, T.A. (2000): Continuous measurement, quantum trajectories, and decoherent histories. Phys. Rev. **A61**, 042107 – [128]

Brune, M., Hagley, E., Dreyer, J., Maître, X., Maali, Y., Wunderlich, C., Raimond, J.M., and Haroche, S. (1996): Observing the progressive decoherence of the "meter" in a quantum measurement. Phys. Rev. Lett. **77**, 4887 – [112]

Brustein, R., and Veneziano, G. (1994): The graceful exit problem in string cosmology. Phys. Lett. **324**, 429 – [168]

Busch, P., Lahti, P.J., and Mittelstaedt, P. (1991): *The Quantum Theory of Measurement* (Springer) – [134]

Caldeira, A.O., and Leggett, A.J. (1983): Path integral approach to quantum Brownian motion. Physica **121A**, 587 – [105]

Caldeira, A.O., and Leggett, A.J. (1985): Influence of damping on quantum interference: An exactly soluble model. Phys. Rev. **A31**, 1059 – [102]

Caldirola P. (1956): A new model of the classical electron. Nuovo Cim. **3**, Suppl. 2, 297 – [33]

Calzetta, E., and Hu, B.L. (1995): Quantum fluctuations, decoherence of the mean field, and structure formation of the early Universe. Phys. Rev. **D52**, 6770 – [155,177]

Carnap, R. (1963): Autobiography. In: *The Philosophy of Rudolf Carnap*, ed. by Schilpp, P.A. (Library of Living Philosophers) p. 37 – [200]

Carroll, S.M. and Chen, J. (2004): Spontaneous Inflation and the Origin of the Arrow of Time. E-print `hep-th/0410270` – [158]

Casati, G., Chirikov, B.V., and Zhirov, O.V. (2000): How many "arrows of time" do we really need to comprehend statistical laws? Phys. Rev. Lett. **85**, 896 – [57]

Casper, B.B., and Freier, S. (1973): 'Gibbs paradox' paradox. Am. J. Phys. **41**, 509 – [71]

Cassirer, E. (1977): *Zur modernen Physik* (Wiss. Buchges. Darmstadt) – [72]

Chiao, R.Y. (1998): Tunneling times and superluminality: A tutorial. E-print `quant-ph/9811019` – [105]

Christodoulou, D. (1970): Reversible and irreversible transformations in black-hole physics. Phys. Rev. Lett. **25**, 1596 – [145]

Cocke, W.J. (1967): Statistical time symmetry and two-time boundary conditions in physics and cosmology. Phys. Rev. **160**, 1165 – [159]

Conradi, H.D., and Zeh, H.D. (1991): Quantum cosmology as an initial value problem. Phys. Lett. **A154**, 321 – [192]

Craig, D.A. (1996): Observation of the final boundary condition: Extragalactic background radiation and the time symmetry of the universe. Ann. Phys. (N.Y.) **251**, 384 – [159]

Cross, M.C., and Hohenberg, P.C. (1993): Pattern formation outside of equilibrium. Rev. Mod. Phys. **65**, 851 – [82]

Dabrowski, M.P., and Kiefer, C. (1997): Boundary conditions in quantum string cosmology. Phys. Lett. **B397**, 185 – [168]

Damour, Th. (2006): Einstein 1905–2005: His approach to physics. In: *Einstein, 1905–2005*, ed. by Damour, Th., Darrigol, O. Duplantier, B., and Rivasseau, V. (Birkhäuser) – [173]

Davidovich, L., Brune, M., Raimond, J.M., and Haroche, S. (1996): Mesoscopic quantum decoherences in cavity QED: Preparation and decoherence monitoring schemes. Phys. Rev. **A53**, 1295 – [112]

Davies, E.B. (1976): *Quantum Theory of Open Systems* (Academic Press) – [99]

Davies, P.C.W. (1977): *The Physics of Time Asymmetry* (University of California Press) – [V]

Davies, P.C.W. (1984): Inflation in the Universe and time asymmetry. Nature **312**, 524 – [159]

Davies, P.C.W., and Fulling, S.A. (1977): Radiation from moving mirrors and from black holes. Proc. R. Soc. London **A356**, 237 – [152]

Davies, P.C.W., and Twamley, J. (1993): Time-symmetric cosmology and the opacity of the future light cone. Class. Quantum Grav. **10**, 931 – [37,159]

Demers, J.-G., and Kiefer, C. (1996): Decoherence of black holes by Hawking radiation. Phys. Rev. **D53**, 7050 – [148]

Denbigh, K.G. (1981): *Three Concepts of Time* (Springer-Verlag) – [16,78]

Denbigh, K.G., and Denbigh, J.S. (1985): *Entropy in Relation to Incomplete Knowledge* (Cambridge University Press) – [76]

d'Espagnat, B. (1966): An elementary note about mixtures. In: *Preludes in Theoretical Physics*, ed. by De-Shalit, A., Feshbach, H., and van Hove, L. (North Holland) – [96]

d'Espagnat, B. (1976): *Conceptual Foundations of Quantum Mechanics* (Benjamin) – [101]

d'Espagnat, B. (1983): *In Search of Reality* (Springer-Verlag) – [101]

d'Espagnat, B. (1995): *Veiled Reality* (Addison-Wesley) – [134,200]

Deutsch, D. (1997): *The Fabric of Reality* (Penguin) – [108,132]

DeWitt, B.S. (1967): Quantum theory of gravity. I. The canonical theory. Phys. Rev. **160**, 1113 – [178,184,188]

DeWitt, B.S. (1971): The many-universes interpretation of quantum mechanics. In: *Foundations of Quantum Mechanics* (Fermi School of Physics **IL**), ed. by d'Espagnat, B. (Academic Press) – [132]

Diósi, L. (1987): A universal master equation for the gravitational violation of quantum mechanics. Phys. Lett. **A120**, 377 – [144]

Diósi, L. (1988): Continuous quantum measurement and Itô formalism. Phys. Lett. **A129**, 419 – [118]

Diósi, L., Gisin, N., Halliwell, J., and Percival, I.C. (1995): Decoherent histories and quantum state diffusion. Phys. Rev. Lett. **74**, 203 – [128]

Diósi, L., and Kiefer, C. (2000): Robustness and diffusion of pointer states. Phys. Rev. Lett. **85**, 3552 – [118]

Diósi, L., and Lukácz, B. (1994): *Stochastic Evolution of Quantum States in Open Systems and in Measurement Processes* (World Scientific) – [118]

Dirac, P.A.M. (1938): Classical theory of radiating electrons. Proc. Roy. Soc. **A167**, 148 – [31]

Dürr, D., Goldstein, S., and Zanghi, N. (1992): Quantum equilibrium and the origin of uncertainty. J. Stat. Phys. **67**, 843 – [127]

Dukas, H., and Hoffman, B. (1979): *Albert Einstein: The Human Side* (Princeton University Press) – [199]

Dyson, F. (1979): Time without end: Physics and biology in an open universe. Rev. Mod. Phys. **51**, 447 – [84]

Eddington, A.S. (1928): *The Nature of the Physical World* (Cambridge University Press) – [6]

Ehrenfest, P., and Ehrenfest, T. (1911): Begriffliche Grundlagen der statistischen Auffassung in der Mechanik. Encycl. Math. Wiss. **4**, No. 32; English translation: *The Conceptual Foundations of the Statistical Approach in Mechanics* (Cornell University Press, 1959) – [52,203]

Einstein, A. (1949): Autobiographical notes. In: *Albert Einstein, Philosopher-Scientist*, ed. by Schilpp, P.A. (The Library of Living Philosophers) – [6]

Einstein, A., Born, H., and Born, M. (1986): *Briefwechsel 1916–1955* (Ullstein) – [8]

Einstein, A., and Ritz, W. (1909): Zum gegenwärtigen Stand des Strahlungsproblems. Phys. Z. **10**, 323 – [19]

Ellis, J., Mohanty, S., and Nanopoulos, D.V. (1989): Quantum gravity and the collapse of the wave function. Phys. Lett. **B221**, 113 – [144]

Everett, H. (1957): 'Relative state' formulation of quantum mechanics. Rev. Mod. Phys. **29**, 454 – **(WZ)** – [127,132,173]

Favre, C., and Martin, P.A. (1968): Dynamique quantique des systèmes amortis non-markovien. Helv. Phys. Acta **41**, 333 – [99]

Feshbach, H. (1962): A unified theory of nuclear reactions. Ann. Phys. (N.Y.) **19**, 287 – [63]

Feynman, R.P., Leighton, R.B., and Sands, M. (1963): *The Feynman Lectures on Physics* (Addison-Wesley) – [75]

Feynman, R.P., and Vernon Jr., F.L. (1963): The theory of a general quantum system interacting with a linear dissipative system. Ann. Phys. (N.Y.), **24**, 118 – [99,105]

Feynman, R.P. (1965): *The Character of Physical Law* (MIT Press) [82]

Flasch, K. (1993): *Was ist Zeit?* (Klostermann, Frankfurt) – [11]

Fokker, A.D. (1929): Ein invarianter Variationssatz für die Bewegung mehrerer elektrischer Masseteilchen. Z. Phys. **58**, 386 – [35]

Foong, S.K., and Kanno, S. (1994): Proof of Page's conjecture on the average entropy of a subsystem. Phys. Rev. Lett. **72**, 1148 – [98]

Ford, J. (1989): What is chaos, that we should be mindful of it? In: *The New Physics*, ed. by P.C.W. Davies (Cambridge University Press) – [78]

Franck, Ph. (1949): *Einstein* (Vieweg) – [200]

Friedman, J., Morris, M.S., Novikov, I.D., Echeverria, F., Klinkhammer, G., and Thorne, K.S. (1990): Cauchy problem in spacetime with closed time-like curves. Phys. Rev. **D42**, 1915 – [16]

Friedman, J.R., Patel, V., Chen, W., Topygo, S.K., and Lukens, J.E. (2000): Quantum superposition of distinct macroscopic states. Nature **406**, 43 – [108]

Fröhlich, H. (1973): The connection between macro- and microphysics. Riv. del Nuovo Cim. **3**, 490 – [51]

Frolov, V.P., and Novikov, I.D. (1990): Physical effects in wormholes and time machines. Phys. Rev. **D42**, 1057 – [16]

Fuchs, C., and Peres, A. (2000): Quantum theory needs no 'interpretation'. Physics Today **53(3)**, 70 – [124]

Fugmann, W., and Kretzschmar, M. (1991): Classical electromagnetic radiation in noninertial reference frames. Nuovo Cim. **106 B**, 351 – [31]

Gabor, D. (1964): Light and Information. Progr. in Optics **1**, 111 – **(LR)** – [75]

Gal-Or, B. (1974): *Modern Developments in Thermodynamics* (Wiley) – [157]

Gardner, M. (1967): Can time go backwards? Sci. Am. **216**, 98 – [72]

Gell-Mann, M., and Hartle, J.B. (1990): Quantum mechanics in the light of quantum cosmology. In: *Complexity, Entropy and the Physics of Information*, ed. by W.H. Zurek (Addison Wesley) – [128,173]

Gell-Mann, M., and Hartle, J.B. (1993): Classical equations for quantum systems. Phys. Rev. **D47**, 3345 – [128]

Gell-Mann, M., and Hartle, J.B. (1994): Time symmetry and asymmetry in quantum mechanics and quantum cosmology. In: *Physical Origins of the Asymmetry of Time*, ed. by Halliwell, J.J., Perez-Mercader, J., and Zurek, W.H. (Cambridge University Press) – [160]

Gerlach, U.H. (1988): Dynamics and symmetries of a field partitioned by an accelerated frame. Phys. Rev. **D38**, 522 – [151]

Ghirardi, G.C., Rimini, A., and Weber, T. (1986): Unified dynamics for microscopic and macroscopic systems. Phys. Rev. **D34**, 470 – [127]

Ghirardi, G.C., Pearle, Ph., and Rimini, A. (1990): Markov processes in Hilbert space and continuous spontaneous localization of systems of identical particles. Phys. Rev. **A42**, 78 – [118]

Gibbs, J.W., (1902): *Elementary Principles in Statistical Mechanics* (Yale University Press) – [48]

Gisin, N. (1984): Quantum Measurement and Stochastic Processes. Phys. Rev. Lett. **52**, 1657 – [118]

Gisin, N., and Percival, I. (1992): The quantum-state diffusion model applied to open systems. J. Phys. **A25**, 5677 – [118]

Giulini, D. (1995): What is the geometry of superspace? Phys. Rev. **D51**, 5630 – [167,190]

Giulini, D. (1998): The generalized thin-sandwich problem and its local solvability. E-print `gr-qc/9805065` – [162]

Giulini, D., and Kiefer, C. (1994): Wheeler–DeWitt metric and the attractivity of gravity. Phys. Lett. **A193**, 21 – [167,190]

Giulini, D., Kiefer, C., and Zeh, H.D. (1995): Symmetries, superselection rules, and decoherence. Phys. Lett. **A199**, 291 – [109,111,181]

Glansdorff, P., and Prigogine, I. (1971): *Thermodynamic Theory of Structure, Stability and Fluctuation* (Wiley) – [60,79,82]

Gleyzes, S., Kuhr, S., Guerlin, C., Bernu, J., Deléglise, Hoff, U.B., Brune, M., Raimond, J.-M., and Haroche, S. (2006): Observing the quantum jumps of light: birth and death of a photon in a cavity. E-print `quant-ph/0612031` – [113,124]

Gold, T. (1962): The arrow of time. Am. J. Phys. **30**, 403 – [137,156]

Goldsmith, T.H. (2006): What birds see. Sci. Am. **295** (Aug), 50 – [14]

Goldstein (1980): *Classical Mechanics* (Addison-Wesley) – [164]

Gorini, V., Kossakowski, A., and Sudarshan, E.C.G. (1976): Completely positive dynamical semigroups of N-level systems. J. Math. Phys. **17**, 821 – [116]

Gott, J.R., and Li, L.-X. (1997): Can the universe create itself? Phys. Rev. **D58**, 3501 – [143]

Gould, A. (1987): Classical derivation of black hole entropy. Phys. Rev. **D35**, 449 – [147]

Gourgoulkon, E., and Bonazzola, S. (1994): A foundation of the virial theorem in general relativity. Class. Quantum Grav. **11**, 443 – [135]

Grad, H. (1961): The many faces of entropy. Comm. Pure Appl. Math. **XIV**, 323 – [61]

Graham, N. (1970): The Everett interpretation of quantum mechanics. Thesis, Univ. North Carolina, Chapel Hill – [134]

Greenberger, D.M., Horne, M.A., Shimony, A., and Zeilinger, A. (1990): Bell's theorem without inequalities. Am. J. Phys. **58**, 1131 – [96]

Griffiths, R.B. (1984): Consistent histories and the interpretation of quantum mechanics. J. Stat. Phys. **36**, 219 – [128]

Groot, S.R. de, and Mazur, P. (1962): *Non-Equilibrium Thermodynamics* (North-Holland) – [66]

Habib, S., Kluger, Y., Mottola, and Paz, J.P. (1996): Dissipation and decoherence in mean field theory. Phys. Rev. Lett. **76**, 4660 – [111]

Habib, S. Shizume, K., and Zurek, W.H. (1998): Decoherence, chaos, and the correspondence principle. Phys. Rev. Lett. **80**, 4361 – [91]

Hájíček, P., and Kiefer, C. (2001): Singularity avoidance by collapsing shells in quantum gravity. Int. J. Mod. Phys. **10**, 775 – [197]

Haken, H. (1978): *Synergetics* (Springer-Verlag) – [82]

Halliwell, J.J. (1989): Decoherence in quantum cosmology. Phys. Rev. **D39**, 2912 – [191]

Halliwell, J.J., and Hawking, S.W. (1985): Origin of structure in the Universe. Phys. Rev. **D31**, 1777 – [154,167,191]

Hao Bai-Lin (1987): *Direction in Chaos* (World Scientific) – [72]

Haroche, S., and Kleppner, D. (1989): Cavity quantum electrodynamics. Physics Today **42**, 24 – [122]

Haroche, S., and Raimond, J.M. (1996): Quantum computing: Dream or nightmare? Phys. Today **49**, (August) 51 – [108]

Harrison, E.R. (1977): The dark night sky paradox. Am. J. Phys. **45**, 119 – [26]

Hartle, J.B. (1998): Quantum pasts and the utility of history. Phys. Scripta **T76**, 159 – [128]

Hawking, S.W. (1975): Particle creation by black holes. Comm. Math. Phys. **43**, 199 – [146]

Hawking, S.W. (1976): Black holes and thermodynamics. Phys. Rev. **D13**, 191 – [147,148]

Hawking, S.W. (1985): Arrow of time in cosmology. Phys. Rev. **D32**, 2489 – [192]

Hawking, S.W. (1992): Chronology protection conjecture. Phys. Rev. **D46**, 603 – [16]

Hawking, S.W., and Ellis, G.F.R. (1973): *The Large Scale Structure of Space-time* (Cambridge University Press) – [143,145]

Hawking, S.W., and Moss, I.G. (1982): Supercooled phase transitions in the very early universe. Phys. Lett. **110B**, 35 – [157]

Hawking, S.W., and Ross, S.F. (1997): Loss of quantum coherence through scattering off virtual black holes. Phys. Rev. **D56**, 6403 – [144]

Hawking, S.W., and Wu, Z.C. (1985): Numerical calculations of minisuperspace cosmological models. Phys. Lett. **151B**, 15 – [167]

Hegerfeldt, G. (1994): Causality problems for Fermi's two-atom system. Phys. Rev. Lett. **72**, 596 – [123]

Heisenberg, W. (1956): Die Entwicklung der Deutung der Quantentheorie. Phys. Blätter **12**, 289 – [85]

Heisenberg, W. (1957): Quantum theory of fields and elementary particles. Rev. Mod. Phys. **29**, 269 – [175]

Hepp, K. (1972): Macroscopic theory of measurement and macroscopic observables. Helv. Phys. Acta **45**, 237 – [89]

Hogarth, J.E. (1962): Cosmological considerations of the absorber theory of radiation. Proc. Roy. Soc. **A267**, 365 – [27]

Holland, P.R. (1993): The de Broglie–Bohm theory of motion and quantum field theory. Phys. Rep. **224**, 95 – [127]

't Hooft, G. (1990): The black hole interpretation of string theory. Nucl. Phys. **B335**, 138 – [194]

Hornberger, K., Hackermüller, L., and Arndt, M. (2005): Influence of molecular temperature on the coherence of fullerenes in a near-field interferometer. Phys. Rev. **A71**, 023601 – [104,106]

Horwich, P. (1987): *Asymmetries in Time* (MIT Press) – [18]

Hoyle, F., and Narlikar, J.V. (1995): Cosmology and action-at-a-distance electrodynamics. Rev. Mod. Phys. **67**, 113 – [38]

Hu, B.L., Paz, J.P., and Zhang, Y. (1992): Quantum Brownian motion in a general environment: A model for system–field interactions. Phys. Rev. **D45**, 2843 – [105]

Imry, Y. (1997): *Mesoscopic Physics* (Oxford University Press) – [110]

Isham, C.J. (1992): Canonical quantum gravity and the problem of time. In: *Integrable Systems, Quantum Groups, and Quantum Field Theory*, ed. by Ibart, L.A., and Rodrigues, M.A. (Kluwer Academic) – [179]

Israel, W. (1986): Third law of black hole dynamics: A formulation and proof. Phys. Rev. Lett. **57**, 397 – [146]

Jadczyk, A. (1995): On quantum jumps, events and spontaneous localization models. Found. Phys. **25**, 743 – [114]

Jammer, M. (1974): *The Philosophy of Quantum Mechanics* (Wiley) – [82,134,177]

Jancel, R. (1963): *Foundations of Quantum and Classical Statistical Mechanics* (Pergamon Press) – [68,94]

Jauch, J.M. (1968): *Foundations of Quantum Mechanics* (Addison-Wesley) – [89]

Jaynes, E.T. (1957): Information theory and statistical mechanics. Phys. Rev. **106**, 620; **108**, 171 – [68]

Joos, E. (1984): Continuous measurement: Watchdog effect versus golden rule. Phys. Rev. **D29**, 1626 – [62,91,107,124]

Joos, E. (1986): Why do we observe a classical spacetime? Phys. Lett. **A116**, 6 – [112,118]

Joos, E. (1987): Time arrow in quantum theory. J. Non-Equil. Thermodyn. **12**, 27 – [176]

Joos, E., and Zeh, H.D. (1985): The emergence of classical properties through interaction with the environment. Z. Phys. **B59**, 223 – [102,103,104,105,106,107]

Joos, E., Zeh, H.D., Kiefer, C., Giulini, D., Kupsch, J., and Stamatescu, I.-O.(2003): *Decoherence and the Appearance of a Classical World in Quantum Theory* (Springer) – [74,103,104,106,107,111,112,114,118,124,126]

Kac, M. (1959): *Probability and Related Topics in Physical Sciences* (Interscience) – [67,203]

Károlyházy, F., Frenkel, A., and Lukácz, B. (1986): On the possible role of gravity in the reduction of the wave function. In: *Quantum Concepts in Space and Time*, ed. by Penrose, R., and Isham, C.J. (Clarendon Press) – [144]

Kauffman, S.A. (1991): *Origin of Order: Self-Organization and Selection in Evolution* (Oxford University Press) – [80]

Kaup, D.J., and Vitello, A.P. (1974): Solvable quantum cosmological model and the importance of quantizing in a special canonical frame. Phys. Rev. **D9**, 1648 – [181]

Kent, A. (1997): Quantum histories and their implications. Phys. Rev. Lett. **78**, 2874 – [128]

Keski-Vakkuro, E., Lifschytz, G., Mathur, S.D., and Ortiz, M.E. (1995): Breakdown of the semiclassical approximation at the black hole horizon. Phys. Rev. **D51**, 1764 – [194]

Khalfin, L.A. (1958): Contribution to the theory of a quasi-stationary state. Sov. Phys. JETP **6 (33)**, 1053 – [120]

Kiefer, C. (1987): Continuous measurement of mini-superspace variables by higher multipoles. Class. Qu. Gravity **4**, 1369 – [191]

Kiefer, C. (1988): Wave packets in mini-superspace. Phys. Rev. **D38**, 1761 – [185, 187]

Kiefer, C. (1992): Decoherence in quantum electrodynamics and quantum cosmol-·ogy. Phys. Rev. **D46**, 1658 – [111,191]

Kiefer, C. (1998): Interference of two independent sources. Am. J. Phys. **66**, 661 – [112]

Kiefer, C. (1999): Decoherence in situations involving the gravitational field. In: *Decoherence: Theoretical, Experimental, and Conceptual Problems*, ed. by Blanchard, Ph., Giulini, D., Joos, E., Kiefer, C., and Stamatescu, I.-O. (Springer) – [112]

Kiefer, C. (2004): Is there an information loss problem for black holes? In: *Decoherence and Entropy in Complex Systems*, ed. by Elze, H.-T., (Springer) – [112,148,197]   ·

Kiefer, C. (2007): *Quantum Gravity*, 2nd edn., (Clarendon Press, Oxford) – [112,170, 181,193,194]

Kiefer, C., Lohmar, L, Polarski, D., and Starobinsky, A.A. (2006): Pointer   states for primordial fluctuations in inflationary cosmology. E-print astro-ph/0610700 – [177]

Kiefer, C., Müller, R., and Singh, T.P. (1994): Quantum gravity and non-unitarity in black hole evaporation. Mod. Phys. Lett. **A9** 2661 – [112]

Kiefer, C., Polarski, D., and Starobinsky, A.A. (1998): Quantum-to-classical transition for fluctuations in the early universe. Int. J. Mod. Phys. **D7**, 455 – [155,177,191]

Kiefer, C., and Singh T.P. (1991): Quantum gravitational corrections to the functional Schrödinger equation. Phys. Rev. **D44**, 1067 – [190]

Kiefer, C., and Zeh, H.D. (1995): Arrow of time in a recollapsing quantum universe. Phys. Rev. **D51**, 4145 – [195,196]

Kolmogorov, A.N. (1954): On conservation of conditionally-periodic motions for a small charge in Hamilton's function. Dokl. Akad. Nauk. USSR **98**, 525 – [72]

Kraus, K. (1971): General state changes in quantum theory. Ann. Phys. (N.Y.) **64**, 311 – [115,116]

Kübler, O., and Zeh, H.D. (1973): Dynamics of quantum correlations. Ann. Phys. (N.Y.) **76**, 405 – [97,99,110,111,133]

Laguna, P. (2006): The shallow waters of the big bang. E-print gr-qc/0608117 – [182]

Lanczos, C. (1970): *The Variational Principles of Mechanics*, 4th edn. (University of Toronto Press) – [165]

Landau, L.D., and Lifschitz, E.M. (1959): *Fluid Mechanics* (Pergamon) – [60]

Landauer, R. (1961): Irreversibility and heat generation in the computing process. IBM J. Res. Dev. **3**, 183 **(LR)** – [76]

Landauer, R. (1996): The physical nature of information. Phys. Lett. **A217**, 188 – [76]

Lapchinsky, V.G., and Rubakov, V.A. (1979): Canonical quantization of gravity and quantum field theory in curved space-time. Acta Physica Polonica, **B10**, 1041 – [190]

Leff, H.S., and Rex, A.F. (1990): *Maxwell's Demon: Entropy, Information, Computing* (Princeton University Press) – [76,209]

Leggett, A.J. (1980): Macroscopic quantum systems and the quantum theory of measurement. Suppl. Progr. Theor. Phys. **69**, 80 – [102]

Levine, H., Moniz, E., and Sharp, D.H. (1977): Motion of extended charges in classical electrodynamics. Am. J. Phys. **45**, 75 – [34]

Lewis, R.M. (1967): A unifying principle in statistical mechanics. J. Math. Physics **8**, 1449 – [54,59]

Lewis, D. (1986): *Philosophical Papers*, Vol. II (Oxford University Press) – [18]

Li, L.-X., and Gott III, J.R. (1998): Self-consistent vacuum for Misner space and the chronology protection conjecture. Phys. Rev. Lett. **80**, 2980 – [194]

Lindblad, G. (1976): On the generators of quantum mechanical semigroups. Comm. Math. Phys. **48**, 119 – [115,117]

Linde, A. (1979): Phase transitions in gauge theories and cosmology. Rep. Progr. Phys. **42**, 389 – [157]

Lockwood, M. (1996): 'Many minds' interpretation of quantum mechanics. Brit. J. Phil. Sci. **47**, 159 – [132]

London, F., and Bauer, E. (1939): *La théorie de l'observation en méchanique quantique* (Hermann, Paris) **(WZ)** – [130]

Louko, J., and Satz, A. (2006): A new expression for the transition rate of an accelerated particle detector. E-print gr-qc/0611117 – [150]

Lubkin, E. (1978): Entropy of an n-system from its correlation with a k-system. J. Math. Phys. **19**, 1028 – [98]

Lubkin, E. (1987): Keeping the entropy of measurement: Szilard revisited. Intern. J. Theor. Phys. **26**, 523 **(LR)** – [130]

Ludwig, G. (1990): *Die Grundstrukturen einer physikalischen Theorie* (Springer-Verlag) – [115]

Mackey, M.C. (1989): The dynamical origin of increasing entropy. Rev. Mod. Phys. **61**, 981 – [55,60,80]

Maeda, K., Ishibashi, A., and Narita, M. (1998): Chronology protection and non-naked singularities. Class. Quantum Grav. **15**, 1637 – [16]

Mayer, J.E., and Mayer, M.G. (1940): *Statistical Mechanics* (Wiley) – [50]

Mensky, M.B. (1979): Quantum restrictions for continuous observations of an oscillator. Phys. Rev. **D20**, 384 – [105]

Mensky, M.B. (2000): *Quantum Measurements and Decoherence* (Kluwer Academic) – [106]

Mermin, D.N. (1998): Nonlocal character of quantum theory? Am. J. Phys. **66**, 920 – [9]

Misner, C.W., Thorne, K.S., and Wheeler, J.A. (1973): *Gravitation* (Freeman) – [161,169,170,194]

Misra, B. (1978): Nonequilibrium, Lyaponov variables, and ergodic properties of classical systems. Proc. Natl. Acad. USA **75**, 1627 – [60]

Misra, B., and Sudarshan, B.C.G. (1977): The Zeno's paradox in quantum theory. J. Math. Phys. **18**, 756 – [62]

Mittelstaedt, P. (1976): *Der Zeitbegriff in der Physik* (BI Wissenschaftsverlag) – [16]

Moniz, E., and Sharp, D.H. (1977): Radiation reaction in nonrelativistic quantum electrodynamics. Phys. Rev. **15**, 2850 – [33]

Monroe, C., Meekhof, D.M., King, B.E., and Wineland, D.J. (1996): A "Schrödinger cat" superposition state of an atom. Science **272**, 1131 – [112]

Monteoliva, D., and Paz, J.P. (2000): Decoherence and the rate of entropy production in chaotic quantum systems. E-print quant-ph/0007052 – [106]

Mooij, J.E., Orlando, T.P., Levitov, L., van der Wal, C.H., and Lloyd, S. (1999): Josephson Persistent-Current Qubit. Science **285**, 1036 – [108]

220   References

Morris, M.S., Thorne, K.S., and Yurtsever, U. (1988): Wormholes, time machines, and the weak energy condition. Phys. Rev. Lett. **61**, 1446 – [16]

Mott, N.F. (1929): The wave mechanics of α-ray tracks. Proc. Roy. Soc. London **A126**, 79 – [104,193]

Mould, R.A. (1964): Internal absorption properties of accelerated detectors. Ann. Phys. **27**, 1 – [31,150]

Nagourney, W. Sandberg, J., and Dehmelt, H. (1986): Shelved optical electron amplifier: Observation of quantum jumps. Phys. Rev. Lett. **56**, 2797 – [113]

Nambu, Y., and Jona-Lasinio, G. (1961): Dynamical model of elementary particles based on an analogy with superconductivity. Phys. Rev. **122**, 345 – [175]

Neumann, J. von (1932): *Mathematische Grundlagen der Quantenmechanik* (Springer-Verlag); Engl. transl. (1955): *Mathematical Foundations of Quantum Mechanics* (Princeton University Press) – [99]

Nicklaus, M., and Hasselbach, F. (1993): Wien filter: A wave-packet-shifting device for restoring longitudinal coherence in a charged-matter-wave interferometer. Phys. Rev. **A48**, 152 – [109]

Nicolai, H., Peeters, K., and Zmaklar, M. (2005): Loop quantum gravity: an outside view. Class Quantum Grav. **22**, R193. – [178]

Norton, J. (1989): Coordinates and covariance: Einstein's view of space-time and the modern view. Found. Phys. **19**, 1215 – [13,163]

Norton, J. (1995): Mach's principle before Einstein. In: *Mach's Principle: From Newton's Bucket to Quantum Gravity*, ed. by Barbour, J.B., and Pfister, H. (Birkhäuser, Basel) – [13]

Omnès, R. (1988): Logical reformulation of quantum mechanics. J. Stat. Phys. **53**, 893 – [128]

Omnès, R. (1992): Consistent interpretation of quantum mechanics. Rev. Mod. Phys. **64**, 339 – [128]

Omnès, R. (1997): General theory of decoherence effect in quantum mechanics. Phys. Rev. **A56**, 3383 – [105]

Omnès, R. (1998): Are there unsolved problems in the interpretation of Quantum Mechanics? In: *Open Systems and Measurements in Relativistic Quantum Theory*, ed. by Breuer, H.P., and Petruccione, F. – [128]

Page, D.N. (1980): Is black-hole evaporation predictable? Phys. Rev. Lett. **44**, 301 – [148]

Page, D.N. (1985): Will entropy increase if the Universe recollapses? Phys. Rev. **D32**, 2496 – [192]

Page, D.N. (1993): Average entropy of a subsystem. Phys. Rev. Lett. **71**, 1291 – [98]

Page, D.N., and Geilker, C.D. (1982): An experimental test of quantum gravity. In: *Quantum Structure of Space and Time*, ed. by Duff, M.J., and Isham, C.J. (Cambridge University Press) – [177]

Page, D.N., and Wootters, W.K. (1983): Evolution without evolution: Dynamics described by stationary observables. Phys. Rev. **D27**, 2885 – [180]

Partridge, R.B. (1973): Absorber theory of radiation and the future of the universe. Nature **244**, 263 – [38]

Pauli, W. (1928): Über das *H*-Theorem vom Anwachsen der Entropie vom Standpunkt der neuen Quantenmechanik. In: *Probleme der Modernen Physik, Sommerfeld-Festschrift* (Hirzel) – [88,90]

Paz, J.P., and Zurek, W.H. (1999): Quantum limit of decoherence: Environment-induced superselection of energy eigenstates. Phys. Rev. Lett. **82**, 5181 – [113]

Pearle, Ph. (1976): Reduction of the state vector by a non-linear Schrödinger equation. Phys. Rev. **D13**, 857 – [118,127]

Pearle, Ph. (1979): Toward explaining why events occur. Int. J. Theor. Phys. **48**, 489 – [99]

Pearle, Ph. (1989): Combining stochastic dynamical state-vector reduction with spontaneous localization. Phys. Rev. **A39**, 2277 – [118]

Pegg, D.T., Loudon, R., and Knight, P.L. (1986): Correlations in light emitted by three-level atoms. Phys. Rev. **A33**, 4085. – [113]

Peierls, R. (1985): Observation in Quantum Mechanics and the collapse of the wave function. In: *Symposium on the Foundations of Modern Physics*, ed. by Lahti, P., and Mittelstaedt, P. (World Scientific) – [100]

Penrose, O., and Percival, I.C. (1962): The direction of time. Proc. Phys. Soc. London **79**, 605 – [59]

Penrose, R. (1969): Gravitational collapse: The role of general relativity. Nuovo Cim. **1**, special number, 252 – [16,138,145]

Penrose, R. (1979): Singularities and time-asymmetry. In: *General Relativity*, ed. by Hawking, S.W., and Israel, W. (Cambridge University Press) – [131]

Penrose, R. (1981): Time asymmetry and quantum gravity. In: *Quantum Gravity 2*, ed. by Isham, C.J., Penrose, R., and Sciama, D.W. (Clarendon Press) – [138,156]

Penrose, R. (1986): Gravity and state vector reduction. In: *Quantum Concepts in Space and Time*, ed. by Penrose, R., and Isham, C.J. (Clarendon Press) – [144]

Percival, I.C. (1997): Atom interferometry, spacetime and reality. Phys. World **10**, (March) 43 – [144]

Peres, A. (1962): On the Cauchy problem in general relativity II. Nuovo Cim. **26**, 53 – [189]

Peres, A. (1980a): Nonexponential decay law, Ann. Phys. (N.Y.) **129**, 33 – [120,122]

Peres, A. (1980b): Measurement of time by quantum clocks. Am. J. Phys. **48**, 552 – [180]

Peres, A. (1996): Separability criterion for density matrices. Phys. Rev. Lett. **77**, 1413 – [98]

Petzold, J. (1959): Zum Anfangswertproblem zerfallender Zustände. Z. Phys. **157**, 122 – [120,122]

Poincaré, H. (1902): *La science et l'hypothèse* (Flammarion) – [12,83]

Pöppel, E., Schill, K., and von Steinbüchel, N. (1990): Sensory integration within temporally neutral system states. Naturwiss. **77**, 89 – [14]

Popper, K. (1950): Indeterminism in quantum physics and in classical physics. British J. Philos. of Science **1**, 117-173 – [72]

Popper, K. (1956): The arrow of time. Nature **177**, 538 – [38]

Price, H. (1996): *Time's Arrow & Archimedes' Point: A View from Nowhen* (Oxford University Press) – [6,8,18,38,42,60,156,199]

Prigogine, I. (1962): *Non-Equilibrium Statistical Mechanics* (Wiley) – [52]

Prigogine, I. (1980): *From Being to Becoming* (Freeman) – [6,72,119,201]

Primas, H. (1983): *Chemistry, Quantum Mechanics and Reductionism* (Springer-Verlag) – [106]

Qadir, A., and Wheeler, J.A. (1985): York's cosmic time versus proper time as relevant to changes in the dimensionless "constants", K-Meson decay, and the unity of black hole and big crunch. In: *From SU(3) to Gravity*, ed. by Grothmann, E.S., and Tauber, G. (Cambridge University Press) – [144]

Qadir, A. (1988): Black holes in closed universes. In: *Proceedings of the Fifth Marcel Grossmann Meeting*, ed. by D. Blair and M. Buckingham (World Scientific) – [169]

Reichenbach, H. (1956): *The Direction of Time* (University of California Press) – [16,18]

Rempe, G., Walther, H., and Klein, N. (1987): Observation of quantum collapse and revival in a one-atom maser. Phys. Rev. Lett. **58**, 353 – [113,122]

Rhim, W.-K., Pines, A., and Waugh, J.S. (1971): Time-reversal experiments in dipolar-coupled spin systems. Phys. Rev. **B3**, 684 – [57]

Rohrlich, F. (1965): *Classical Charged Particles* (Addison-Wesley) – [29,32,34]

Rohrlich, F. (1963): The principle of equivalence. Ann. Phys. (N.Y.) **22**, 169 – [31]

Rohrlich, F. (1997): The dynamics of a charged sphere and the electron. Am. J. Phys. **65**, 1051 – [34]

Rohrlich, F. (2001): The correct equation of motion for a classical point charge. Phys. Lett. **A 283**, 276 – [34]

Röpke, G. (1987): *Statistische Mechanik für das Nichtgleichgewicht* (Physik-Verlag, Weinheim) – [66]

Rovelli, C., and Smolin, L. (1990): Loop representation for quantum general relativity. Nucl. Phys. **B133**, 80 – [178]

Ryan, M.P. (1972): *Hamiltonian Cosmology* (Springer) – [187]

Saunders, S. (2005): On the Explanation of Quantum Statistics. E-print quant-ph/0511136 – [49]

Sauter, Th., Neuhauser, W., Blatt, R., and Toschek, P.E. (1986): Observation of quantum jumps. Phys. Rev. Lett. **57**, 1696 – [113]

Schlögl, F. (1966): Zur statistischen Theorie der Entropieproduktion in nicht abgeschlossenen Systemen. Z. Phys. **191**, 81 – [70]

Schloßhauer, M. (2004): Decoherence, the measurement problem, and interpretations of quantum mechanics. Rev. Mod. Phys. **76**, 1267 – [134]

Schloßhauer, M. (2006): Experimental motivation and empirical consistency in minimal no-collapse quantum mechanics. Ann. Phys. (N.Y.) **321**, 112 – [103]

Schmidt, E. (1907): Zur Theorie der linearen und nichtlinearen Integralgleichungen. Math. Ann. **63**, 433 – [97]

Schrödinger, E. (1935): Discussion of probability relations between separated systems. Proc. Cambridge Phil. Soc. **31**, 555 – [97]

Schulman, L.S. (1997): *Time's Arrows and Quantum Mechanics* (Cambridge University Press) – [156,159]

Schulman, L.S. (1999): Opposite thermodynamical arrows of time. Phys. Rev. Lett. **83**, 5419 – [57]

Schuster, H.G. (1984): *Deterministic Chaos* (Physik-Verlag, Weinheim) – [72]

Scully, M.O., Englert, B.-G., and Walther, H. (1991): Quantum optical tests of complementarity. Nature **351**, 111 – [105]

Shannon, C. (1948): A mathematical theory of communication. Bell System Techn. J. **27**, 370 – [68]

Shor, P.W. (1994): Algorithms for Quantum Computation: Discrete Logarithms and Factoring. In: *35th Annual Symposium on Foundations of Computer Science*, ed. by S. Goldwasser (IEEE Computer Society Press) – [108]

Smart, J.J.C. (1967): Time. In: *Encyclopedia of Philosophy*, Vol. 8, ed. by Edwards, Th. (Free Press) – [8]

Smoluchowski, M. Ritter von (1912): Experimentell nachweisbare, der üblichen Thermodynamik widersprechende Molekularphänomene. Phys. Z. **13**, 1069 – [73]

Sokolovsky, D. (1998): From Feynman histories to observables. Phys. Rev. **A57**, R1469 – [132]

Sonnentag, P., and Hasselbach, F. (2005): Decoherence of Electron Waves Due to Induced Charges Moving Through a Nearby Resistive Material. Braz. J. Phys. **35**, 385 – [110]

Spohn, H. (2000): The critical manifold of the Lorentz-Dirac equation. Europh. Lett. **50**, 287 [34]

Squires, E.J. (1990): *Conscious Mind in the Physical World*, (Hilger) – [132]

Szilard, L. (1929): Über die Entropieverminderung in einem thermodynamischen System bei Eingriffen intelligenter Wesen. Z. Phys. **53**, 840 – (**WZ,LR**) – [73]

Taylor, J.H. (1994): Binary pulsars and relativistic gravity. Rev. Mod. Phys. **66**, 711 – [28]

Tegmark, M. (1993): Apparent wave function collapse caused by scattering. Found. Phys. Lett. **6**, 571 – [118]

Tegmark, M. (1998): The interpretation of quantum mechanics: Many worlds or many words. Fortschr. Physik **45**, 855 – [173,201]

Tegmark, M. (2000): Importance of quantum decoherence in brain processes. Phys. Rev. **61**, 4194 – [107,128,174]

Tegmark, M., Aguirre, A., Rees, M.J., and Wilczek, F. (2006): Dimensionless constants, cosmology and other dark matters. Phys. Rev. **D73**, 023505 – [84]

Teitelboim, C. (1970): Splitting of the Maxwell tensor: Radiation reaction without advanced fields. Phys. Rev. **D1**, 1572 – [30]

Thiemann, T. (2006a): Solving the problem of time in general relativity and cosmology with phantoms and k-essence. Preprint `astro-ph/0607380` – [163]

Thiemann, T. (2006b): Loop quantum gravity: An inside view. `hep-th/0608210` – [178]

Timpson, C.G. (2005): The Grammar of Teleportation. E-print quant-ph/0509048 – [96]

Tolman, R.C. (1938): *The Principles of Statistical Mechanics* (Oxford University Press) – [40]

Ulfbeck, O., and Bohr, A. (2001): Genuine Fortuitousness: Where did that click come from? Found. Phys. **31**, 757 – [114,201]

Unruh, W.G. (1976): Notes on black-hole evaporation. Phys. Rev. **D14**, 870 – [150,152]

Unruh, W.G., and Wald, R.M. (1982): Acceleration radiation and the generalized second law of thermodynamics. Phys. Rev. **D25**, 942 – [148,153]

Vaihinger, H. (1911): *Die Philosophie des Als Ob* (Scientia Verlag) – [83]

Van Hove, L. (1957): The approach to equilibrium in quantum statistics. Physica **23**, 441 – [92]

Van Kampen, N.G. (1954): Quantum statistics of irreversible processes. Physica **20**, 603 – [91]

Veneziano, G. (1991): Scale factor duality for classical and quantum strings. Phys. Lett. **B265**, 287 – [168]

Vilenkin, A. (1989): Interpretation of the wave function of the universe. Phys. Rev. **D39**, 1116 – [191]

Wald, R.M. (1980): Quantum gravity and time reversibility. Phys. Rev. **D21**, 2742 – [144]

224   References

Wald, R.M. (1997): Gravitational collapse and cosmic censorship. E-print gr-qc/9710068 – [143]
Wehrl, A. (1978): General properties of entropy. Rev. Mod. Phys. **50**, 221 – [52]
Weinberg, S. (1977): *The First Three Minutes* (Basic Books, New York) – [171]
Weizsäcker, C.F. von (1939): Der Zweite Hauptsatz und der Unterschied zwischen Vergangenheit und Zukunft. Ann. Physik **36**, 275 – [82]
Weizsäcker, C.F. von (1982): *Die Einheit der Natur* (Hanser, München) – [6,201]
Werner, R.F. (1989): Quantum states with Einstein–Rosen correlations admitting a hidden variable model. Phys. Rev. **A40**, 4277 – [98]
Wheeler, J.A. (1979): Frontiers of time. In: *Problems in the Foundations of Physics* (Fermi School of Physics **LXXII**), ed. by G.T. di Francia (North Holland) – [72]
Wheeler, J.A. (1994): Time today. In: *Physical Origins of the Asymmetry of Time*, ed. by Halliwell, J.J., Perez-Mercader, J., and Zurek, W.H. (Cambridge University Press) – [201]
Wheeler, J.A., and Feynman, R.P. (1945): Interaction with the absorber as the mechanism of radiation. Rev. Mod. Phys. **17**, 157 – [24,36,38]
Wheeler, J.A., and Feynman, R.P. (1949): Classical electrodynamics in terms of direct interparticle action. Rev. Mod. Phys. **21**, 425 – [24,38]
Wheeler, J.A., and Zurek, W.H. (1983): *Quantum Theory and Measurement* (Princeton University Press) – [209]
Whitrow, G.J. (1980): *The Natural Philosophy of Time* (Clarendon Press) – [16]
Wick, G.C., Wightman, A.S., and Wigner, E.P. (1952): The intrinsic parity of elementary particles. Phys. Rev. **88**, 101 – [89]
Wigner, E.P. (1972): Events, laws of nature, and invariance principles. In *Nobel Lectures – Physics* (Elsevier) p. 248 – [1]
Wilkinson, S.R., Barucha, C.F., Fischer, M.C., Madison, K.W., Morrow, P.R., Niu, Q., Sundaram, B., and Raizen, M.G. (1997): Experimental evidence for non-exponential decay in quantum tunneling. Nature **387**, 575 – [122]
Williams, D.C. (1951): The myth of the passage. J. Philosophy **48**, 457 – [12]
Willis, C.R., and Picard, R.H. (1974): Time-dependent projection-operator approach to coupled systems. Phys. Rev. **A9**, 1343 – [59]
Winter, R.G. (1961): Evolution of quasi-stationary state. Phys. Rev. **123**, 1503 – [122]
Witten, E. (1997): Duality, spacetime and quantum mechanics. Phys. Today **50**, (May) 28 – [178]
Woolley, R.G. (1986): Molecular shapes and molecular structures. Chem. Phys. Lett. **125**, 200 – [106]
Wootters, W.K. (1984): "Time" replaced by quantum correlations. Int. J. Theor. Phys. **23**, 701 – [180]
Yaghjian, A.D. (1992): *Relativistic Dynamics of a Charged Sphere* (Springer), App. D – [33]
Yoccoz, J.-C. (1992): Travaux de Hermann sur les tores invariantes. Asterisque **206**, 311 – [55]
York, J.W. (1983): Dynamical origin of black-hole radiance. Phys. Rev. **D28**, 2929 – [146]
Zeh, H.D. (1967): Symmetrieverletzende Modellzustände und kollektive Bewegungen. Z. Phys. **202**, 38 – [181]
Zeh, H.D. (1970): On the interpretation of measurement in quantum theory. Found. Phys. **1**, 69 – (**WZ**) – [102,103,106,132,132,173]

Zeh, H.D. (1971): On the irreversibility of time and observation in quantum theory. In: *Foundations of Quantum Mechanics* (Fermi School of Physics **IL**), ed. by d'Espagnat, B. (Academic Press) – [102]

Zeh, H.D. (1973): Toward a quantum theory of observation. Found. Phys. **3**, 109 – [102,132,133]

Zeh, H.D. (1979): Quantum theory and time asymmetry. Found. Phys. **9**, 803 – [132]

Zeh, H.D. (1983): Einstein nonlocality, space-time structure, and thermodynamics. In: *Old and New Questions in Physics, Cosmology, Philosophy, and Theoretical Biology*, ed. by van der Merwe, A. (Plenum) – [169,197]

Zeh, H.D. (1984): *Die Physik der Zeitrichtung* (Springer) – [V,179]

Zeh, H.D. (1986a): Die Suche nach dem Urzeitpfeil. Phys. Blätter **42**, 80 – [176]

Zeh, H.D. (1986b): Emergence of classical time from a universal wave function. Phys. Lett. **A116**, 9 – [179,191]

Zeh, H.D. (1988): Time in quantum gravity. Phys. Lett. **A126**, 311 – [183]

Zeh, H.D. (1993): There are no quantum jumps, nor are there particles! Phys. Lett. **A172**, 189 – [113]

Zeh, H.D. (1994): Time-(a)symmetry in a recollapsing quantum universe. In: *Physical Origins of the Asymmetry of Time*, ed. by Halliwell, J.J., Perez-Mercader, J., and Zurek, W.H. (Cambridge University Press) – [192,196]

Zeh, H.D. (1999a): Note on the time reversal asymmetry of equations of motion. Found. Phys. Lett. **12**, 193 – [34]

Zeh, H.D. (1999b): Why Bohm's quantum theory? Found. Phys. Lett. **12**, 197 – [127,193]

Zeh, H.D. (2000): Conscious observation in quantum mechanical description. Found. Phys. Lett. **13**, 221 – [128,132]

Zeh, H.D. (2003): There is no "first" quantization. Phys. Lett. **A309**, 329 – [49,93, 125,172]

Zeh, H.D. (2004): The wave function: It or bit? In: *Science and Ultimate Reality*, ed. by Barrow, J.D., Davies, P.C.W., and Harper, C.L. (Cambridge University Press) – quant-ph/0204088 – [124,134]

Zeh, H.D. (2005a): Where has all the information gone? Phys. Lett. **A347**, 1 – [112,144,169,197]

Zeh, H.D. (2005b): Remarks on the compatibility of opposite arrows of time. Entropy **7**, 199 (http://www.mdpi.org/entropy/list05.htm#issue4) – [57,159]

Zeh, H.D. (2005c): Roots and fruits of decoherence. E-print quant-ph/0512078 – [96,103,194]

Zurek, W.H. (1981): Pointer basis of quantum apparatus: Into what mixture does the wave packet collapse? Phys. Rev. **D24**, 1516 – [102]

Zurek, W.H. (1982a): Environment-induced superselection rules. Phys. Rev. **D26**, 1862 – [102,103]

Zurek, W.H. (1982b): Entropy evaporated by a black hole. Phys. Rev. Lett. **49**, 1683 – [147]

Zurek, W.H. (1986): Reduction of the wave packet: How long does it take? In: *Frontiers of Nonequilibrium Statistical Physics*, ed. by G.T. Moore and M.T. Scully (Plenum) – [103]

Zurek, W.H. (1989): Algorithmic randomness and physical entropy. Phys. Rev. **A40**, 4731 – [72]

Zurek, W.H. (1991): Decoherence and the transition from quantum to classical. Phys. Today **44**, (Oct.) 36 – [105]

Zurek, W.H., Habib, S., and Paz., J.P. (1993): Coherent states via decoherence. Phys. Rev. Lett. **70**, 1187 – [107,111]

Zurek, W.H., and Page, D.N. (1984): Black-hole thermodynamics and singular solutions of the Tolman–Oppenheimer–Volkhoff equation. Phys. Rev. **D29**, 628 – [147]

Zurek, W.H: and Paz, J.P. (1994): Decoherence, chaos, and the second law. Phys. Rev. Lett. **72**, 2508 – [106]

Zwanzig, R. (1960): Statistical mechanics of irreversibility. Boulder Lectures in Theoretical Physics **3**, 106 – [58]

Index

Words of Acclaim for this Fifth Edition

This book is a real gem, providing concise yet clear and comprehensive coverage of the classic problem of time, ranging from classical electromagnetism and thermodynamics to quantum cosmology. The discussion is lucid and intuitive without glossing over the important details. It is a treat to learn about the role of decoherence from the man who discovered it.

Max Tegmark

H. Dieter Zeh, one of the main founders of decoherence theory, sets forth here a most penetrating analysis of recent advances in classical statistical mechanics, quantum physics and theoretical cosmology, enabling him to convincingly establish the observer-relatedness of the concept of entropy, along with that of macroscopic description in general. Both, he shows, are ultimately due to the unavoidable use of the relevance notion, itself linked to the fact that we, observers, are local whereas whatever may be called objective (i.e. mind-independent) reality is unquestionably nonlocal.
Certainly one of the most informative, thorough and thought-provoking books on the subject.

Bernard d'Espagnat

Dieter Zeh has made important pioneering contributions to the interpretation of quantum mechanics and the study of the direction of time. His comprehensive and frequently republished book on the fascinating issue of time's arrow has few if any rivals.

Julian Barbour

This is a great book. It engages the reader in a philosophical meditation on time (much clearer, and in my opinion deeper than Heidegger's "Being and Time"). It is moreover a learned book in which every necessary physical or mathematical argument is given explicitly and cleanly. I learned much from it and nothing that I happened to know was missing.

Roland Omnès

[...] an authoritative treatment of an exciting and controversial subject: The origins and implications of the arrow of time are traced from its genesis at the microscopic quantum level to the macroscopic thermodynamic setting and include the analysis of Maxwell's demon. The monograph concludes with a provocative discussion of the arrow of time in the context of quantum gravity and cosmology. Dieter Zeh guides the reader with a sure hand through the meanderings of the subject, pointing out unexpected connections, and shedding new light on the remaining mysteries.

Wojciech Zurek